Atari Age

Atari Age

The Emergence of Video Games in America

Michael Z. Newman

The MIT Press
Cambridge, Massachusetts
London, England

This book was set in Stone Sans and Stone Serif by Toppan Best-set Premedia Limited. Printed and bound in the United States of America.

Library of Congress Cataloging-in-Publication Data

Names: Newman, Michael Z., author.
Title: Atari age : the emergence of video games in America / Michael Z. Newman.
Description: Cambridge, MA : MIT Press, [2017] | Includes bibliographical references and index.
Identifiers: LCCN 2016028476 | ISBN 9780262035712 (hardcover : alk. paper)
Subjects: LCSH: Video games--United States. | Video games industry--United States.
Classification: LCC GV1469.3 .N484 2017 | DDC 794.8--dc23 LC record available at https://lccn.loc.gov/2016028476

10   9   8   7   6   5   4   3   2   1

Contents

Acknowledgments

So many people have helped me produce this book and I am grateful for many kinds of support from institutions, friends, family, and even people on the Internet I barely know.

For sharing their ideas, suggestions, research, or even just the name of someone else who might know the answer to a question, thank you Megan Ankerson, Catherine Baker, Anthony Bleach, Will Brooker, Rachel Donohoe, Christine Evans, Kevin Ferguson, Raiford Guins, Thomas Haigh, Carly Kocurek, Melanie Kohnen, David McGrady, Stuart Moulthrop, Sheila Murphy, Laine Nooney, Rebecca Onion, Tommy Rousse, Phil Sewell, Kent Smith, Colin Tait, Jacqueline Vickery, Ira Wagman, and Mark J. P. Wolf. Anonymous readers for the MIT Press offered outstanding feedback.

I am so happy to have found communities of scholars on Facebook and Twitter who answer questions and give advice. On Tumblr, I am thrilled to follow hundreds of people I do not know "in real life" who share images, videos, GIFs, and links. Even if I don't know you personally, your presence in my networks enriches my knowledge and experience every day.

Thanks are also due to a number of institutions and people who serve them. Ellen Engseth and other University of Wisconsin-Milwaukee archivists helped me get my hands on a treasure trove of department store catalogs. The Interlibrary Loan office of my campus library is doing God's work, and I owe them more than I can offer here. UWM's Center for 21st Century Studies and its former director, Richard Grusin, were essential in helping me shape this project when it was getting started and giving me time to work on it. I could not have completed this work without an Arts & Humanities Travel Grant and a Graduate School Research Committee Award, and I am grateful for those forms of support. Librarians, archivists, and support staff at the International Center for the History of Electronic Games at The Strong, UCLA Film and Television Archive, and the Library of Congress aided me in many ways. Thanks in particular are due to J. P. Dyson, Thomas

Hawco, and Lauren Sodano of The Strong/ICHEG, and Mark Quigley of UCLA.

At the MIT Press it has been my pleasure to work with Susan Buckley, Susan Clark, Judy Feldmann, Pamela Quick, and Doug Sery.

I want to acknowledge some of the sources of information that we all rely on and tend not to cite in our scholarly publications: Google Books, Google Scholar, Amazon "look inside," and Wikipedia. I use these constantly to look things up. I often go to them even when sources I need are on the bookcase next to my desk or stored on the hard drive of my computer. Wikipedia in particular is so useful because so many volunteer editors have given generously of their time and knowledge, and anyone who ever wants facts quickly owes them their thanks.

In spring 2012 I taught a seminar on video games to graduate students, and I learned an amazing amount from its participants. Stephen Kohlmann, Alexander Marquardt, Pavel Mitov, Max Neibaur, Carey Peck, Leslie Peckham, and David Wooten, thanks for all of your contributions to our collective understanding of games and their history.

My colleagues in the Department of Journalism, Advertising, and Media Studies at the University of Wisconsin-Milwaukee are supportive in many ways. I want in particular to acknowledge the generosity of David Allen, Rick Popp, Jeff Smith, and Marc Tasman. Elana Levine is a wonderfully helpful colleague and spouse.

I presented portions of this book as work in progress to audiences at UWM; Marquette University; more than one Console-ing Passions International Conference on Television, Video, Audio and New Media; conferences of the Society for Cinema and Media Studies and the American Studies Association; the Fun with Dick and Jane: Gender and Childhood conference at the University of Notre Dame; and the Interplay conference at Northwestern University and the University of Chicago. Thanks to all who organized these conferences and in particular to the Interplay conference participants and organizers, including Reem Hilu. Thanks to my audiences for your attention and your questions and feedback.

Allan Zuckerman and Ron Becker were sources of old game consoles and cartridges. Some of these were also passed down from the collection of my late father-in-law Elliott Levine.

I am grateful to every friend and acquaintance who told me where in their childhood home the video games would be found. I also want to acknowledge my childhood friends and friends of friends in whose basements I played Atari, Intellivision, and Colecovision as a child, and with whom I went off by bicycle, bus, or subway to Toronto's public spaces of play.

My mother-in-law Dodie Levine's basement on Woodview Lane in Park Ridge, Illinois, was an inspiration to me. In the years when I visited it, this space contained two 1970s pinball machines, a one-armed bandit slot machine, a ping-pong table, a personal computer, an upright piano terribly out of tune, a well-stocked bar, and many sundry hobbyist and collector artifacts. I often reflected on the status of public amusements in the home while playing with my children down there, and thought of that room as a time machine to the 1970s. Research happens in the library and the archive, but it also happens during moments of everyday life when we encounter people, objects, and spaces who prompt us to think and reflect and wonder.

Anyone who inquired about my video games book and how it's going, or asked me what I'm working on, maybe you were just making conversation—I appreciate it. You gave me opportunities to encapsulate my ideas and offered a sense of how the world would receive them.

Many thanks are due to Leo Newman, not only a dear son but a research assistant and companion in play, and his brother Noah Newman, equally dear, barely a toddler when I started this work and as of this writing, the only member of the family who really appreciates the animated TV series *Pac-Man*. No one has helped me more and in as many ways as Elana Levine, my partner in so many things. In addition to commenting on chapter drafts, taking unnecessary words out of sentences, and sharing sources, she has sustained me and our family while I have been at work, and given me the inspiration of her own scholarly example. Such great gratitude is owed to my family, friends, and networks, who made this work better, and indeed, made it possible. Thank you all.

Preface

Video games have been part of my life since my childhood, but I have found myself intensely interested in them during two periods: the early 1980s, and the years I have spent on this book.

I began the research for *Atari Age* not out of any particular desire for recapturing the past, but out of an interest in one aspect of the history of television. While writing about digital television innovations of the early twenty-first century, such as DVRs and online video, I wanted to understand a longer history of TV's technological improvements. Ideas about video games in the 1970s, along with ideas about cable TV, videotape cassette recorders, and other new ways of using a TV set, were remarkably similar to ideas about television during the era of digital convergence. In particular, people assumed that TV was in need of a technological upgrade to give its viewers more agency and alleviate problems associated with mass media.[1] This book began as a project of tracing the history of interactive moving-image technologies, of entwining video game and television history. After all, the medium's name includes a word that was for many years a synonym for TV, and "video games" was used interchangeably in the 1970s with "TV games."

While looking into this connection, I discovered that relatively little had been written about early video games, particularly little social and cultural history of the medium as it emerged, and I became eager to help fill that gap.[2] Early cinema and early television had been studied in illuminating and influential historical work.[3] Early video games, I thought, had the potential to be just as productive for historical study. I also saw that doing research on this topic would give me an excuse to read old magazines, which I knew would be fun. This book is the result.

Once immersed in research, it was probably unavoidable that I would come to an expanded understanding not only of games and related media, but also of my own childhood. Although it wasn't my conscious agenda,

I did relive my younger years through writing this. I was born in 1972, the same year as the debut of *Pong*, and I was ten years old in 1982, when the most desirable plaything in North America was an Atari 2600. I played video games including *Tron*, *Tempest*, and *Ms. Pac-Man* (as well as pinball games) at Uptown Variety, a convenience store on Eglinton Avenue I could ride my bike to from my family's home in a residential neighborhood of Toronto. I played console games in the rec rooms of friends' houses. I also played in arcades like the clean and respectable Video Invasion on Bathurst Street, where classmates had birthday parties, and some seedier, less wholesome spots on Yonge Street downtown, where I traveled by subway. I owned a few handheld electronic games, but we had no Atari console (or any of its rivals), and the family's first PC arrived years later. My own parents were suspicious of video games and objected to their presence in our home. They shared a fear (which I discuss in chapter 5) with many other grown-ups at the time: that their children's success in school and their childhood development could be threatened if they became addicted to playing Atari games all the time. As a result, I have always regarded video games, like many kinds of popular media, as something of a forbidden pleasure at odds with adults and their culture. This feeling lingers even now, when I have the Atari I so desperately wanted as a child.

I cannot really feel in middle age what I might have as a child. Some early games are disappointingly dull or confusing to me now, though I love the candy-colored stripes of *Super Breakout* and the abstract splotches of *Asteroids*. Happily, the pleasures of clearing *Ms. Pac-Man* mazes of pellets and devouring blue ghosts have not faded. Writing this became an effort to substitute for my lost childhood thrills an intellectual pleasure of producing insight and preserving the meanings of the past through their historical interpretation. My mom and dad's refusal must, at least in some tiny way, be an origin point for this book, so in retrospect I can be grateful for the deprivation. It has helped me to appreciate in a personal way what was so appealing and also so worrisome about video games when they were new.

Introduction: Early Video Games and New Media History

Since the turn of the twenty-first century, a lively topic in cultural studies and within the culture industries has been *new media*, a term that has never been simple to define.[1] Rather than just whatever media happen to be new at a particular time, new media usually have something more specifically to do with computers as media of communication. New media are distinguished historically from print, film, broadcasting, and other mainstays of modern life, which are relegated to the status of old media. New media are seen as the opposite of the twentieth century's mass media, since they are participatory rather than passive. Video games, as popular interactive technology of the digital age, are a key example.

Talk of new media often functions as hype within the world of business, where digital technologies have been seen as catalysts for disruptive innovation, but intellectuals are also susceptible to overheated excitement. Such rhetoric can err by seeing the present as an implausibly radical break from the past. This way of looking at changing technology tends to obscure as much as it reveals, promoting (or decrying) present events as fundamental departures rather than recognizing continuities with the past. At the same time, historically minded writers have recognized that new media studies could be taken up as a paradigm for doing media history, going against the grain of the sometimes frenzied optimism or pessimism of new media rhetoric.[2] Whether digital or not, any medium was new once, and most tend to be renewed over time as material changes prompt fresh ways of using and thinking about a technology. New media begin in a period of mysterious uncertainty and potential, a period of becoming, but eventually they are integrated into markets, regulatory frameworks, and the patterns of everyday life. Over time, they come to seem familiar and unremarkable. A medium's period of buzzy novelty can be particularly important for establishing its meaning and value. These often abide many decades or centuries after newness has passed. The long history of a medium is shaped (though not in all ways) by early understandings and uses.

This book is a new media history of video games, charting their emergence in the United States during their first decade of public, commercial availability, beginning around 1972. That year marked the release of the first home console, the Magnavox Odyssey, and Atari's first hit coin-operated game, *Pong*. Video games had been in the works for some years by this time, but 1972 was the moment when they arrived as an experience of play wide open to American consumers. As a new media history, *Atari Age* assumes that during their emergence, video games were objects without fixed meanings, without a clear identity, without a commonly shared understanding of their cultural status. All of this had to be worked out. The emergence of video games was not just a matter of bringing new products to market for people to buy and use. It was not merely a succession of platforms, interfaces, and games. It was also a process of introducing ideas, including notions of who should play games, where and when, and for what purposes. It was not clear in 1972 whether games would be seen as a harmless amusement or as a danger to America's children. It was not clear if their players would mainly be adults or kids, or of one or another gender. It was not clear if the medium would be seen in positive, productive terms, or rather as a moral or physical threat. Ideas like these are up for grabs when a medium is new.

A history of emerging media is one of uncertainties and misdirections, of struggles over uses and purposes, of unexpected and surprising outcomes. Inventors cannot dictate how the fruits of their labor will be understood and appreciated, who will use them and to what ends. Technologies pass through a period of interpretive flexibility, as different social groups adopt them for divergent purposes, before a process of closure establishes a clearer identity, making some uses dominant while others become more marginal or are cast aside.[3] What makes early video games distinct from games in later periods in the history of the medium is precisely this lack of a stable identity.

When video games were new, people apprehended their novelty through associations with already familiar technologies and experiences. Just as automobiles were called horseless carriages, video games were familiarized by comparison with existing objects, next to which they were often regarded as improvements. New media, as Jay David Bolter and Richard Grusin argue, are *remediations*, adapting and repurposing the forms and techniques of earlier media.[4] The older forms through which video games established an identity varied in reputation and cultural status. Pinball had a sketchy life as a public amusement sometimes banned for its associations with gambling and crime, but also adored by countercultural fans who admired the rebellious image of the pinball player. Television had been regarded as a promising medium for democratic civic life that became debased by commercialism and catering to the mass audience to the point where it was loathed for inculcating passivity and disengagement. Computers were seen alternately as instruments of dehumanization employed by massive institutions of corporate or state control, and as miraculous technologies that promised to solve myriad problems. Games played in the home, such as board or card games, were associated with the suburban family ideal and the ideology of domestic harmony so important to the postwar consensus society.

Popular imagination closely linked early video games with these very different technologies, media, and social practices, though it sometimes also distinguished them. In some ways, video games were caught between these technologies and practices. Each of these remediations was also centrally concerned with age, gender, and class identity politics, as each of the "old" media carried along social identities of its own coming from the spaces and users with which it was identified. As a new form of public amusement, video games picked up associations with the history of coin-operated games, but also contrasted against the earlier versions of coin-op machines. As a new thing to play in the home, video games were part of a history of domestic leisure, but also were seen as a more masculine amusement than the typical family-room game. As a device to plug into a television set, video games intervened in television history by giving broadcast audiences an alternative to watching TV shows. As a use for home computers, video games were an incentive to some early PC users to acquire a computer, but also provided a seemingly trivial reason to own one, wasting expensive cutting-edge technology. Public amusements, family leisure, television, and computers all pushed and pulled video games one way or another in the formulation of their cultural status as a medium.

As they emerged, video games became associated with players of certain identities, and the consequences of this are still with us. Decades after their emergence, gender inequality suffuses the world of video games. Despite the presence of huge numbers of girls and women as players, video games cater especially to boys and men in many ways, including the representations within them. The culture of video games often seems dominated not just by boys and men, in ways that exclude girls and women, but by strongly identified gamers who seem threatened by any form of critique of their fandom and pastime.⁵ Women are underrepresented in the games industry, an industry that can be inhospitable to them. When women speak out about unequal representation and other gender inequities in the world of video games, they are often harassed, threatened, and vilified. This book ends its story around 1983, but its ideas about how video games became associated with masculinity (along with youth and middle-class identity) provide a backstory to much more recent developments in the history of the medium. Many key ideas about games familiar to people in the 2010s were circulating already in the 1970s and 80s. This book is, among other things, a history of how video games became masculinized during their period of emergence.

It is also a history of how video games became identified with two other aspects of social identity: age and class. In their first decade, games came to be regarded as a somewhat respectable type of boy culture: as a medium for male, middle-class players in their preteen, teenage, and early adult years. This was a contrast against two conceptions of leisure that also shaped the history of electronic games, which informed their promotion and reception. The first was the ideal of the companionate family at play in the bourgeois suburban home with members of different ages and sexes. The second was the reputation of public coin-operated amusements long associated with gambling, crime, sex, drugs, and riffraff, a reputation shaped by lower-class identity and by a more mature masculinity.

The youthful, masculine, and middle-class status of the medium was a product of many dynamics and influences. Unlike many new media, video games emerged as multiple objects in different kinds of spaces. They were both a computer and a television technology. They were both a public and private amusement, played in taverns and living rooms. In some ways they were like pinball or pool, and in others they were like watching television or playing checkers. The status and identity that emerged for video games was a product of negotiations among many meanings and values. It developed along with changes in public amusement spaces, computers in everyday life, and the family. The identification of the medium with an identity of its user was not merely a product of who played most often, though this was part of it. It was also a matter of the place the medium occupied in popular imagination. It was a product of how people looked at and thought about video games.

This book is as much concerned with ideas and representations as with game consoles and cartridges, but never with one at the expense of the other. One basic assumption will be that these two sides of the history inform one another. They cannot be isolated from each other if we want to know the cultural significance of the emerging medium. Some of our understanding of early video games comes through looking at the games themselves, but just as often, context and surroundings tell the story. Spaces of play and identities of players are no less important than game companies and the products they sold. The pleasures of playing video games and their place in the everyday lives of their players are of no less interest than design innovations over this first decade. This is not a technical or aesthetic history or an economic analysis of an emerging industry as much as a social and cultural study of the medium in relation to its players. In this way, *Atari Age* is at once a work of game studies and media studies, looking at games as games, as play, as a medium of their own, but also as a form of media that has continuities and commonalities with other forms and is approachable using the same tools as other cultural studies of media. Its ambition is to uncover assumptions and expectations about video games that became established as a shared (even if contested) common sense in the period of their emergence, as cultural histories have done for cinema, radio, and TV, among other media.

This common sense did not emerge without struggle. At every point along the way of this story, we find not one clear idea of video games but competing positions in tension and contradiction with one another. There was no one common sense about video games to which everyone learned to subscribe. Rather, video games were perched between rival conceptions, and differences had to be worked out over time, bit by bit. The middle-class status of the new medium developed in tension with a less respectable and less legitimate reputation of public arcades and game rooms. The close linkage between video games and television required a distinction between the promising interactivity of the former and the reputation of the latter as a vast wasteland. The place of video games in the home was inflected by the idea of domestic space as feminine, and by the unified family ideal of private leisure. The use of computers to play games and the status of games as computerized playthings introduced tensions between productive and frivolous uses of advanced electronic technology. The craze for arcade games in public places stoked many adults' fears of young people's corruption, both moral and cognitive, by this new form of amusement even while many experts touted its benefits. This is a book about these tensions and contradictions, which were typically resolved by fashioning the new medium around some identities more than others. The admission of video games into the realm of mainstream popular culture required an accommodation of identities other than young and male in the world of electronic play even as boys were affirmed as the most central segment of the market.

Video Games in the United States, 1972–1983

While this is a book about the early history of video games, it does not share the ambition of some of the video game histories already published. Books such as *Replay: The History of Video Games*, *The Golden Age of Video Games: The Birth of a Multi-Billion Dollar Industry*, and *The Ultimate History of Video Games* offer a particular kind of representation in which the key elements are video game companies and the people (mostly men) who worked for and ran them; technologies and their commercial release as consumer products; and particular games and genres, along with their aesthetic and technical development.[6] These journalistic histories contain much important and useful information, sometimes based on interviews with key figures in the industry, and they function as useful reference works. This book is different. It makes no attempt to cover every console or every commercially successful or aesthetically interesting arcade cabinet. It is mostly concerned with how people understood and thought about video games as a whole.[7]

Thus, *Atari Age* does not chronicle year-by-year the fortunes of the video game business or the release schedule of its products. Such information is easily accessible, but nevertheless a brief encyclopedia-style history, highlighting key moments and objects in the period under discussion, is offered here as an orientation for the reader.

The history of video games struggles to designate a "first" game. It might be William Higinbotham's *Tennis for Two*, or the MIT game *Spacewar!*, or Ralph Baer's Brown Box, the prototype for the Odyssey.[8] The first video game console you could buy to play at home was the Odyssey. The first coin-operated video game many people encountered in a public place was *Pong*, though the much less successful *Computer Space* (1971) preceded it. Computers were used to play games in the 1940s and '50s, but these are rarely named as potential firsts. It actually matters little for a social or cultural history which of these games was the original, or if any game deserves to be so honored. For a small number of people who had access to computers in the workplace or at school, computer games using a graphical output were available for play before 1972, but by and large 1972 was when the general public gained the opportunity to access them.

Following *Pong*'s success, versions of the ball-and-paddle game were released for both the home and arcade.[9] Arcades and other public game spaces (which had existed for decades) contained a variety of games in the early and mid-1970s, including pinball, kiddie rides, and electro-mechanical driving and shooting games, in addition to fully electronic video games. Many *Pong* copycats were sold as home TV games. Some could also be used to play hockey or soccer as ball-and-paddle contests, perhaps in color. It was a crowded field, and the appeal of these games was hard to sustain after the novelty wore off. The popular press covered these games as a new way of using a television set. In the mid-1970s, electronic games were, along with videocassette recorders, among the "new tricks your TV can do."[10]

In the second half of the 1970s, changes in the business, technology, and experience of video games broadened their appeal and improved their com-mercial fortunes. Some of the earliest games, such as *Pong* and the Odyssey, were electronic but not computerized, and contained no microchips or soft-ware. They were made using television technology.[11] The introduction of silicon chips into a host of consumer culture technologies from cash regis-ters and calculators to home computers and toys like Merlin and Simon also transformed video games technologically. The big news in home games in the second half of the decade was *programmable* consoles like Atari's Video Computer System (VCS, later the Atari 2600). A programmable console would accept cartridges with their own chips so that it could play many different games, which increased the utility of the device and expanded its owner's interest in play. Although Atari's VCS was by far the most commer-cially and culturally successful home leisure product of the video game industry, it had many rivals including Fairchild's Channel F, which pre-ceded it to market, and Magnavox's Odyssey^2. Its most serious competition was Mattel's Intellivision, whose games sometimes had more sophisticated graphics, and whose controller was far more complicated than Atari's sim-ple four-direction joystick and fire button.

Late in the decade, *Space Invaders* was released first as an arcade cabinet and then as an Atari cartridge. Like *Pong*, the success of this game in public game rooms promoted the sale of the home version. By the time of *Space Invaders'* success, Atari had been acquired by the media conglomerate Warner Communications, Inc. (WCI), and a few years later Atari was earning more for WCI than film, television, or any other division, and accounted for more than half of the company's operating profits.[12] Many consumers eager to play *Space Invaders* at home bought an Atari console for this very purpose, a dynamic repeated with *Asteroids*, *Missile Command*, and *Pac-Man*.

At the same time as Atari and Intellivision struggled for dominance in the home market and hit games like *Space Invaders* earned large sums in quarters dropped in the coin slot, microcomputers became available to consumers to purchase for the home. In the late 1970s and early 1980s, millions of Americans acquired a computer such as a TRS-80 from Radio Shack, a Commodore PET or VIC-20, an Atari 400 or 800, or an Apple II. While the uses of such machines were presented in marketing discourses as virtually limitless, the most common application of home computers was to play and sometimes to program games. Earlier computers had a variety of inputs and outputs, but what came to be known as a personal computer had a cathode ray tube (CRT) monitor and speakers as its output. Even if they saw other uses as more important, users of early home computers tended to try out playing games on these high-tech devices. Many game programs were sold for PCs, and these were among the most successful software products of the time.

At the end of the 1970s and in the first few years of the 1980s, video games exploded commercially and became a huge cultural sensation. The press now covered them not so much as a novelty to be introduced to an unfamiliar public, but as a newly popular form of leisure that was making some people quite rich, claiming more and more of young people's time and money, and potentially causing harm or bringing benefits to players. Intellectuals weighed in more and more on the significance of this new medium that was suddenly out-earning movies and records, and a fan press sprouted up to feed the interests of newly devoted players of the video generation. New arcade games continued to lead the way forward, with home games lagging behind them in technological sophistication. The arcade was typically considered a more authentic site for playing video games, and home consoles were advertised as recreating the arcade experience. But arcades were also feared by many adults as a place where children might be led astray, and video games more generally were objects of grave concern for some parents and teachers who believed they would have negative effects on habitual players.

The years 1982 and 1983 marked the pinnacle of popularity for early arcade and home video games. *Pac-Man* and *Ms. Pac-Man* had become practically universally appealing, making video games a familiar part of many people's leisure-time experiences. By this time video games were fairly clearly established in popular imagination and their cultural status was worked out. The flexibility of their meanings was being closed off, and the associations between the medium and the typical identity of its players had been set. What followed in the video games industry was a crash, as many products failed, companies went out of business, and the trade faced declining fortunes. What caused the crash was probably, to give the broadest explanation, a mismatch of supply and demand. A glut of products had been brought to market to satisfy a craze for video games, some of them wanting in quality. Many players at home were using computers rather than consoles. In late 1982, Atari's report of lower than expected earnings caused a drop in video game company stock prices and rattled investor confidence.[13] But from the players' perspective, the video games crash was hardly remarkable. The same spaces of play continued to offer video games, and the consoles in the home continued to be played. This history stops in 1983 because the crash is a widely regarded historical milestone, and because an identity for video games as a medium had been established by the time it occurred. But *Atari Age* is a social and cultural history rather than a business or economic one; this will be the last word on the crash.

In each of the phases of video games' first decade, particular problems need to be worked out, and particular questions need to be answered. Who would play games, and where and when? What games would be popular, and why? What value would different games have for different players? Would video games be welcomed or feared, or some combination of these? Would they be seen as productive or problematic, legitimate or a waste of time and money?

Early Games on Their Own Terms

Just as early cinema is different from studio-system Hollywood and early television is different from the height of the three-network era, video games in the 1970s and early 1980s are different in many ways from later video games. This is true both of the games themselves and of the ways people thought about them. It's tempting to look at these games as the first chapter in a longer narrative of video game history, which in some ways they are. But there is a tendency in telling the story of an art form or cultural practice to project backward and see early stirrings as anticipatory of later developments. We could, for instance, see the people who stuffed the *Pong* machine full of quarters at Andy Capp's Tavern in Sunnyvale, California, as the first gamers. We could look at *Battlezone* as an origin point for the first-person shooter, or at the intermissions in *Ms. Pac-Man* as proto-cutscenes. But we ought to be wary of anachronistic thinking. The history of video games does not necessarily lead anywhere, and the period of early video games should be seen on its own terms rather than as the prelude to a later understanding of what the medium is or could be. No one knew what a gamer, a shooter, or a cutscene was in the 1970s and early 1980s because these things didn't exist. The challenge of this kind of history is to project ourselves back into an earlier mindset, to understand the medium from the perspective of its users at the time, and to appreciate the period for what it was.

To see video games, having debuted in 1972 (or before), as part of a continuous history to the present day and beyond, is to risk sidelining much of importance about video games in the period of their emergence. They were not initially regarded as their own independent medium with a clear and distinct identity. They were likely seen as another public or home amusement or another use for a television set or computer. In some instances, they were familiarized as a marriage of television and computer.[14] Video game history, television history, computer history, and the histories of arcades and rec rooms are not neatly distinct from one another. In particular, the early history of a medium will overlap significantly with earlier media—so radio is essential for understanding the emergence of television, traditions of live theater and performance are essential for understanding the emergence of cinema, and telegraphy and telephony are essential for understanding the emergence of radio. Games are no different.

In addition to taking on the identity of neighboring media or overlapping with their functions and pleasures, early periods in the history of a medium tend to excite or frighten a public uncertain about its status. New media are conceptualized in some strikingly consistent ways in disparate historical periods, conjuring up similar notions of technology auguring society's redemption or ruin. The same dreams of democratic participation and the same worries about private life being eradicated and public life being trivialized have attended emerging new media across the generations.[15]

Early video games, like other emerging media, were an occasion for hopes and fears not only about the medium but perhaps more importantly about the society into which it was emerging. New media predictably excite the public about their potential, sometimes in positive and even utopian terms, and other times provoking dystopian reactions that express widely felt anxieties.[16] The desire for video games to overcome television's deficiencies or to teach young people computing was not only an expression of hope for the new medium but also an opportunity to work out society's issues with mass media's effects on social life, childhood development in a changing world, and economic transformations linked to technological change. The fears of video games corrupting youth or ruining their minds likewise had as much to do with uncertainties around raising children in a culture perceived to be threatening to them and the tensions that always exist between generations. New media, early media, predictably inspire these divergent reactions. Many of the same fantasies that the Internet conjured up in the late twentieth and early twenty-first centuries were also once inspired by the telegraph, telephone, cinema, radio, television, and video games.

These are all examples of media of communication and representation, some of the key information technologies of modernity. By referring to each of these as a medium, I mean to include both their material and conceptual qualities. As in my book *Video Revolutions: On the History of a Medium*, I take a cultural view of the concept of medium. A medium is made up of both things and ideas, and both of these influence each other and change over time. Video games comprise circuitry and its connections and containers, input interfaces, displays and speakers, packaging, art and design, but also commonly shared notions of what they are and what and whom they are for. A medium is defined not just by the parts and the ways they work, the things they can or cannot do, but also by a social identity, a cultural status, a degree of legitimacy and respectability. For instance, cinema and television have an identity related to but also distinct from film cameras, projectors, and the transmission and reception of pictures across electromagnetic waves, with each having its own cultural status relational to the other. These more social or cultural dimensions of a medium are products not only of technology as such but, crucially, of lived social relations of power that place the medium in popular imagination by identifying it with some users and not others, some purposes and not others, some ideals and not others.[17]

The identity of the emerging medium of video games that this book is concerned with is not innocent of power relations, but is rather shot through with their implications. Video games became youthful, masculine, and middle class not by accident but through the negotiation of their identity in relation to those earlier media against or alongside which they were understood. The place of computers, television, coin-op machines, rec room games, and other pertinent media within these social relations, and each one's cultural status, is central to video games' emerging meanings and values. Video games were defined through these comparative discourses, where their meanings were constructed.

These meanings may not have been shared universally, and the idea of a "popular imagination" may emphasize common meanings at the expense of peculiar or minority visions. But like words and their definitions, cultural concepts often have broad purchase among members of a society. Even if there are doubters and dissenters among the public, or people who just don't get the message, there are also dominant, commonsense assumptions that inform popular imagination, and the task of this historical work is to apprehend them.

Archives and Sources of Early Game History

How do we access the meanings a medium or technology had in the past? How can historical research give us an understanding of what people thought about video games when they were new, and what place they might have had in the experiences of their players? How can we know now what their meanings were, what value they had, and for whom?

Looking at the games themselves helps, but it doesn't tell us everything we might want to know. The meanings and ideas around a medium are discursive: they circulate in many places as forms of knowledge that are widely shared. Knowledge about video games comes from games as material objects, but also from a diverse array of discourses in which games are discussed, debated, promoted, perhaps denigrated or celebrated, and more generally represented as objects and experiences with particular affordances, for particular users. The sources of knowledge about video games that this book has drawn from include materials produced by the business itself, such as packaging and catalogs, and advertising and promotional texts such as television commercials. It also draws on similar marketing discourses such as Sears Christmas Wish Books.

The sources of this history also include several overlapping categories of the press: popular newspapers and magazines for general or intellectual readers, periodicals for those interested in business in particular, the trade presses of the electronics, retail, and amusement industries, and publications aimed at video game fans. Depending on their readership, these print sources position video games in certain ways, making sense of them for readers. Sometimes these sources are introducing a new media technology to the public, and sometimes they are aimed at workers in the amusements trade seeking to maximize their profits. Sometimes they are covering video games as a trade or as a form of entertainment. By the early 1980s, many publications had started up to cash in on the craze for video games by selling magazines and books to mostly young male video game enthusiasts.

Contemporaneous social science research is another source of knowledge about video games, which is both evidence of who played and in what ways, but also of how the medium was understood at the time by the researchers themselves. Some of this writing is based on quantitative or qualitative survey research, and some of it is ethnographic. The fields range across the social sciences: sociologists, psychologists, and education researchers were particularly interested in games. Several books by intellectuals in these fields published in the early 1980s shed light on how some elite thinkers conceived of the medium, including David Sudnow's *Pilgrim in the Microworld*, Sherry Turkle's *The Second Self*, and Loftus and Loftus's *Mind at Play*.[18]

Representations of games in media, as in films or television programs such as *What's My Line?* (the Odyssey appears as a mystery guest), *Airplane!* (air traffic controllers play an Atari game on the radar display), and *WarGames* (a teenage boy instigates a nuclear crisis by playing what he believes to be an online game), also shed light on popular conceptions of the medium. Games appear in a variety of media texts of the 1970s and '80s as a novelty or new social force. They can be both positive and negative influences on the typically young players who take interest in them. The ideas expressed in these representations are evidence of widely circulating conceptions of the medium.

None of these sources speaks for itself. All are produced to advance particular interests, and all represent a point of view and perhaps an agenda of positioning the medium in a certain way. But they are all traces left behind that show the ways of understanding and imagining video games available at the time, from which ordinary people would have drawn their own interpretations and understandings. This kind of research assumes that ideas about popular culture, although not universal or compulsory, tend to be broadly shared discourses that are both produced and reflected by popular media. These sources, when synthesized and situated in historical context, establish a horizon of expectations against which people at the time made sense of the social world.[19] Such sources, as Lynn Spigel argues, "do not reflect society directly" and offer no straightforward evidence of "what people do, think, or feel." But they can be read "for evidence of what they read, watch, and say," and by showing us these things, popular media help us tap into a history of ordinary people's fantasies and pleasures.[20] The emerging identity of games can be understood by reading these sources and putting them into conversation with one another. The identity of video games was a product of many forces and developments in these years. It becomes visible through the interpretation of games, representations of games, and ideas circulating about players and play with electronic amusements.

In the historiography of fairly recent popular culture, what counts as sources, and where are the archives that house them? Video game history is blessed to have both legions of living fans and an open Internet on which those fans share an amazing array of documents, images, and videos. Many games themselves are emulated online for anyone to play. YouTube alone is an archive of video game history with an impressive enough collection of materials to sustain many research projects. The Internet Archive contains full scans of issues of *Electronic Games* and *Byte* among its numerous magazine offerings. Of course not everything is accessible online, and this book's research also finds sources in conventional archives including the International Center for the History of Electronic Games, the Vanderbilt Television News Archive, and the UCLA Film and Television Archive. Libraries also function as archives, and periodicals on microfilm or in bound volumes on the stacks, as well as scans provided by interlibrary loan, are essential documentary evidence for this study. Some materials were also acquired in ways that might seem unconventional, like shopping on eBay and accepting Atari and Intellivision cartridge donations from the collections of friends and family, but there is no logical reason to observe any old-fashioned distinction between more and less official or legitimate sources of historical knowledge. The history of popular media and everyday life cannot be written without materials that may be archived casually or unintentionally, and we must access them however we can, remaining critical of their status as sources while also deriving meaning from them. We all have archives of popular media and everyday life in our possession and sources of historical knowledge can be constituted as such by regarding these treasures or ephemera as sources, as evidence of the past. This is a matter of how to look at objects rather than some quality inherent in them or in their consecration in official archival institutions. Each historical researcher working on popular media amasses his or her own archive, a combination of official historical documents such as legal decisions and newspaper stories and other sources some traditional historians might consider odd or out-of-bounds. Any source that speaks of the history of video games in everyday life is welcome in my archive.

Preview

The chapters that follow explore the emergence of video games in both public places like arcades and private spaces of the home, beginning with the origins of the video arcade and ending with Pac-Man Fever. Each chapter is centered on a cultural tension or contradiction through which the emerging identity of the new medium was worked out.

Chapter 1, "Good Clean Fun: The Origins of the Video Arcade," charts the course of public coin-operated games and other amusements in the twentieth century leading up to the shift from pinball to video games as the most profitable and popular form of public play, drawing from the trade press of the coin-op amusements business, particularly *RePlay*. During the 1970s, many suburban game rooms were fashioned as "family fun centers," assuring them of middle-class respectability. This occurred before the rise to dominance of video games, when pinball was still the most important source of income for the coin-op trade. By riding pinball's wave of new-found respectability in these more upscale and culturally legitimate spaces, video games found a welcome spot in which to appeal to middle-class young people. But video games, which did not carry along pinball's associations with gambling and crime, also improved the reputation of arcades by being technologically sophisticated, and by being deemed acceptable for home play within more affluent American neighborhoods.

Chapter 2, "'Don't Watch TV Tonight. Play It!' Early Video Games and Television," reveals the common threads of the histories of video games and television. Early games were typically framed as a form of television or a use for a television set. News items and promotional discourses used the language of broadcasting, for instance, telling audiences unfamiliar with electronic play that you tune in the game like any other channel. Names like Intellivision and Channel F reference TV and related concepts. But video games were presented as improvements on TV, ways of solving the older medium's putative problems of passivity and low cultural value, according to the terms of the midcentury mass society critique. By presenting video games as participatory, champions of the new medium showed their potential to redeem television from its status as a plug-in drug. This discourse of rehabilitation of TV has an undercurrent of gender and class politics, as a feminized mass medium, long associated with passivity, was transformed by a more active and masculinized technological marvel.

Chapter 3, "Space Invaders: Masculine Play in the Media Room," explores the significance of video games as domestic amusements, placing the new medium in the space of the idealized suburban home. The new games were often presented in marketing and advertising materials as a way to bring the family together during times of domestic leisure. Commercials, game brochures, store catalogs, and photos in magazines pictured players of mixed ages and both genders enjoying each other's company through electronic play. This effort to sell the new medium to families emphasized an inclusive gender and class appeal. But the forms and genres of the games themselves, and many advertisements, present a rather contradictory appeal to young boys in particular, emphasizing youth and masculinity. Games were seen not only as a way of unifying families, but also as a means of escape for boys from a feminized space, continuing a long tradition of boy culture but moving it indoors.

Chapter 4, "Video Games as Computers, Computers as Toys," locates the emergence of video games alongside the development of home computers in the later 1970s, showing these two histories to be mutually entwined. Computers had been used to program games for decades before *Pong* and the Odyssey, and before Apple and Atari. Games could demonstrate the abilities of computers and impressed ordinary observers of computing. They were also often used to familiarize novices with computers, most typically middle-class boys and men. When home computers became a consumer good, games were among their most frequent uses, though this was often seen as a wasteful use of the technology. For children in particular, games could teach computing, but playing with computers itself was a central appeal of the home computer revolution. Drawing on advertisements, trade press discourses, and publications aimed at home computer users, this chapter argues that early video games and home computers share a history, that play was crucial to the development of PC culture, and that computers gave cultural legitimacy to the emerging medium of video games.

Chapter 5, "Video Kids Endangered and Improved," pairs two related developments that occupied many Americans' attention in the early 1980s. On the one hand was a panic about video games and young people, particularly young boys spending quarters playing arcade games, which was a frequent topic of news stories when municipalities took steps to regulate and in some instances ban coin-operated video games. On the other hand was an effort to counter this hysteria, particularly among social scientists and other experts, who saw in video games a great potential benefit to young people. In addition to teaching eye-hand coordination, video games were seen as training in technology that would be essential to professional knowledge work in the postindustrial society. This chapter argues that these fearful and optimistic ideas about video games were two sides of the same coin, which expressed a certainty that the incredible popularity of this new medium would have profound and lasting effects on its users, particularly those who were young, male, and middle class.

Finally, chapter 6, "Pac-Man Fever," takes up the most popular early video game and the period around 1982–1983 when games became a huge pop culture sensation. *Pac-Man* was an unlikely game to become so phe-nomenally popular as, unlike *Space Invaders, Asteroids, Defender*, and many other hits, it had no spaceships and no shooting. *Pac-Man* was cute and cartoonish, and its difference (along with its appeal as a challenging and fun game) helped it attract a wider market of players than other games, particularly girls and women, for whom it had been designed. This chapter concludes the book by considering the importance of this blockbuster game as an exception to the medium's close identification with boy culture. It argues that *Pac-Man* and its sequel *Ms. Pac-Man* broadened the medium's appeal and acceptability but also, by being so exceptional, reinforced the identification of video games with masculinity.

Throughout the decade of video games' emergence, we find clashing values and meanings, divergent ideals of the new medium's purpose and function. The pages that follow reveal the negotiation of an identity for video games between competing conceptions of players and their experiences.

1 Good Clean Fun: The Origins of the Video Arcade

By the early 1980s, a new space had emerged in American cities, suburbs, and towns. The video arcade was an amusement center or game room found in retail spaces on busy streets and inside shopping centers, bowling alleys, miniature golf courses, eateries, beachside boardwalks, hotels, bus depots, and airports. A video arcade, noisy with electronic beeps and futuristic computer music, would contain a few dozen upright cabinets lined up against a wall at which mainly young, male players would stand dropping quarters into the slot, maneuvering spaceships and creatures around a screen with a joystick or trackball and firing at foes by rhythmically pressing on buttons. More young, mostly male patrons of the arcade would often stand at the player's side as spectators or rivals, observing, judging, learning, anticipating, admiring. The most popular games in the arcade were huge hits for the coin-operated amusements business, which had many decades of history by this point. Each *Asteroids*, *Zaxxon*, *Centipede*, or *Ms. Pac-Man* cabinet had the potential to earn thousands of dollars a year for the operator, who placed the cabinets in public locations, and the proprietor, who shared the income from the cash box. Video games were more lucrative than any jukebox, pool table, or pinball machine ever had been. They were also a veritable sensation, a novel form of high-tech youth culture perceived to pose a moral threat but also at the same time to promise a newly enriched society of electronically mediated pleasure and work.[1]

In addition to the arcades, in the 1970s and 1980s, video games alone or in small clusters were installed in supermarkets and 7 Elevens, coin laundries, movie theater lobbies, college student unions, and taverns. Anywhere people might have a bit of extra time and spare change in their pockets was a good location for a video upright, which sometimes took a spot once occupied by a pinball machine. Within a few years, this new kind of technologically advanced play was introduced into a wide variety of public places. A generation born in the 1960s and '70s took to these games as something of an obsession, just as earlier generations had grown up going to the pictures or listening to rock and roll. Coin-operated video game cabinets were essential to the leisure of teenagers, and particularly teenage boys, coming of age in the last years of the Cold War.[2]

Beginning with *Computer Space* in 1971, and *Pong* and the Odyssey, both in 1972, electronic TV games were available for home and arcade play. Some games were also played in workplaces or universities on mainframes and minicomputers. The space of the video arcade did not define the emerging medium all by itself. But the arcade was in some ways more important than home, work, or school: it led the way in establishing popular games like *Space Invaders* and *Pac-Man*. It also set in place associations between video games and a history of public amusements, making the home game into a version of something else, an adaptation of arcade play for domestic space. Home video game consoles were marketed as a way of recreating the arcade experience, but with the perceived threats of the game rooms removed by a shift from public to private.

These threats were well publicized.[3] News reports addressed fears of children's lunch money being squandered on coin-operated amusement, of schoolwork abandoned for play, of addiction comparable to getting hooked on hard drugs, of minds ruined by overabsorption in electronic microworlds.[4] The United States Surgeon General spoke publicly saying video games could be harmful to children and cause them to be violent, and although he quickly clarified that his assertions were based on no scientific evidence, the news of his concern spread widely.[5] Arcades were feared as dens of crime and depravity, and many municipalities tried to regulate them out of existence. But the idea that coin-operated amusements were a threatening form of popular culture was not first expressed in the 1980s. Such fear was the product of many decades of concern about public amusements. It was expressed by early twentieth-century progressives, moralizing crusaders against vice, and grandstanding public servants. Coin-operated amusements were represented in popular media from the 1930s to the 1970s as a diversion for lowlifes and a trade under the control of criminals with crooked politicians in their pockets. The video arcade—and the meanings associated with it arising around spaces of electronic pleasure—emerged as a product of both established and shifting values and associations with a history stretching back to the nineteenth-century American city.

These meanings stitch together a number of forms of pleasure-seeking that span better than a century of public life. The video arcade of the 1980s inherits a tradition of arcades (or playlands, sportlands, and whatever else amusement rooms have been called) and of the machines found in them. The emergence of video games was not just the birth of a new kind of object and a new culture of play. It was also the continuation of established practices and ideas about spaces of amusement, and the renewal of coin-operated leisure with new technologies and fresh modes of experience. This history encompasses a number of diverse and overlapping forms of media, games, and consumer culture, including motion pictures, musical recordings, sports, carnival attractions, gambling, and games of chance and skill. It has at different places and times been seen as urban and suburban, seedy and classy, more and less culturally legitimate. But the history of public coin-operated amusements is largely one of wanting legitimacy and struggling for respectability. The success of the video arcade is, in part, the winning of this legitimacy and respectability. At the same time, though, video games carried on old connotations of disrepute, and occasioned a new moral panic around young people and media. The arcade of the 1970s and '80s was seen in multiple, contradictory ways.

And the improved reputation of amusement centers was hardly caused by video games alone. Video games were able to occupy a central place in popular culture so quickly in part by emerging into a public space that existed already. Amusement arcades began as penny arcades in the 1890s, and passed through a number of iterations over the decades, as different kinds of amusement went in and out of fashion. In the 1970s their image was transformed by suburbanization, new business practices, and the rehabilitation of pinball as "good clean fun." The newly respectable arcades of the 1970s were not dominated by video games, which were present as a minority among a variety of types of game but played second fiddle to pinball. Pinball was a great force for the transformation of public coin-operated amusements in the early years of video games, and video games rode pinball's coattails until the very end of the decade, when one video game changed the trajectory of public amusement sharply. Before the fantastic success of *Space Invaders* at the end of the decade, pinball was king. Video games did not cause arcades to change their reputation and place in popular imagination very radically at all, but rather nudged them farther along a path they were already traveling. Video games picked up the meanings of these newly respectable spaces, while they also renewed the threat historically occasioned by youth cultures centered around public amusements.

This perceived danger was a product of widely shared associations between spaces and identities. Public coin-operated amusements were never consecrated as high culture, with its associations of an affluent, adult, contemplative experience. They were rather lively and boisterous, cheap and common, accessible to practically any member of society. Young people, working class and mostly male, were their best patrons over many years (except in drinking establishments, where children were excluded). The video arcade became a more middle-class version of this kind of space, but even so it retained a strong residual character of earlier sites of play, a sense of the tough masculine culture of which polite society disapproves. The emergence of video games as a medium was a product of these associations between public play spaces and a culture of boys and young men engaged in competitive action.

Coin-Ops before *Pong*

Coin-in-the-slot machines, while not an invention of the nineteenth century, became quite common in its later years with the spread of urban consumer culture, vending such inexpensive items as cigars and razor blades. Often called slot machines or simply slots, the machines would replace human labor in many fields as symbols of the advancing machine age. One such slot machine supplanted the "try-your-weight apparatus" previously staffed by boy attendants.[6] Similar coin-in-the-slot testing and guessing games filled amusement parks in the nineteenth and twentieth centuries alongside fortune-tellers and merry-go-rounds.[7] The many dime museums in American cities of the nineteenth century contained penny arcades with displays of curios and freaks and stages for dramatic performances. The coin-in-the-slot amusements in a dime museum in the 1880s or '90s would have included various trial-test machines and "Cosmic and Dioramic views," or peepshows. A dime museum was a place of affordable amusement (ten cents to enter, when a vaudeville show might cost a quarter) combining lectures or theatrical performance with other attractions, and having a more proletarian character than the legitimate stage, World's Fair exposition, or fine arts museum with which it coexisted. Along New York's Bowery, dime museums operated in shops on streets where one would also find cheap melodrama theaters and penny arcades.[8]

By comparison to the dime museum, the arcade was free to enter and even cheaper in reputation. It was noted for being "the cheapest kind of amusement ever concocted for the delectation of an audience of countrymen and boys bound on seeing the sights of the city and seeing them cheap. In the penny arcade the low water mark of cheapness has been reached. Nothing could be cheaper."[9] In contrast to rivals for young people's spare change in the early twentieth century, such as candy shops and ice cream parlors, a moral question surrounded the content of the amusement in the penny arcade.[10] Around the turn of the century, the penny arcade in an American city would have contained a variety of attractions including punching bags, automatic scales, shooting rifles, fortune tellers, and phonograph players. The most popular coin-op amusement of the 1890s and 1900s, however, lined up in long rows, was the moving picture peepshow machine such as a mutoscope. It contained a succession of images printed on cards, which would become animated when made to flip before the viewer's eye by hand-cranking the device. Progressives disparaged the views for sale through these coin-op movie machines as "stupid" even if "tame," though they were also known to contain such so-called vulgar imagery as women in states of undress.[11] Some of the titles of peepshow pictures popular in the 1890s indicate these subjects: *Parisienne Girls*, *Dressing Room Scene*, *Getting Ready for the Bath*, and *Little Egypt*, the last being a famous hootchie-cootchie dancer. The hand-crank gave the user control over the speed of the picture, perhaps allowing the viewer to slow down at a moment of voyeuristic pleasure.[12] The sexual content of the mutoscope was not only found in the picture cards. The highly suggestive posters advertising the mutoscopes were often more cause for protest than the motion-picture images. One critic called these posted advertisements "the most undesirable features of an arcade."[13] The pictures themselves were also condemned by polite observers as vile and scandalous, filling young boys' minds with "evil and debasing thoughts."[14] They were "the lowest and meanest trick yet devised for snatching away the pennies and the morals of the people."[15]

Coin-slot recordings of music had been popular beginning around 1890 in midways, resort areas, and railroad stations, but also in urban parlors dedicated to them, which also contained vending machines.[16] But the penny arcade enjoyed its brief heyday beginning when photographic moving pictures emerged later in the decade (the mutoscope's debut was in 1897) and ending once theaters dedicated to projecting them on a screen became the standard way of experiencing cinema, in the later 1900s.[17] Once a program of films lasting thirty or more minutes could be seen for five or ten cents in a Nickelodeon theater, there was little demand for peepshows charging one cent for forty to sixty seconds.[18] Coin-operated moving picture peepshows were a popular diversion in many kinds of places, including theater lobbies, which continued to house coin-op machines when the featured attraction was a projected film. Amusement rooms were located on busy streets in entertainment districts like New York City's 14th Street, which housed the Automatic One-Cent Vaudeville of Adolph Zukor, who would proceed from there eventually to become a Hollywood mogul. Many of these rooms were in resort areas, like New York's Coney Island and countless other resorts within an easy mass transit ride from an American city. The establishment of the penny arcade as an amusement center free to enter containing coin-operated entertainment of brief duration would be maintained through several generations of public play and leisure even as the fortunes of penny arcades rose and fell. But once motion pictures attained the status of a maturing industry with well-established institutions of production, distribution, and exhibition, the mutoscopes were relegated to nostalgic novelty, and the penny arcade sought other attractions. Zukor and his ilk dropped the coin business when cinema became more theater and less carnival. Penny arcades endured in part as a token of the simpler, happier times of years past, as in the penny arcade of Disneyland's small-town simulation, Main Street USA. By the 1960s the coin-operated machines of the first arcade boom were presented as museum pieces.[19]

While Nickelodeon theaters were often held to be dangerous for young and female patrons unchaperoned in their dark, stranger-filled auditoriums, they might also have been regarded as improving on penny arcades, which were not only peepshow parlors to satisfy men's baser desires, but also places for tough street children to hang around without even spending one penny.[20] Their cheapness and questionable clientele gave coin-operated amusements a low reputation that would endure long after motion pictures had ceased to be their main attraction. This was only exacerbated when both the arcade and the coin-operated amusement device enjoyed a resurgence in the 1920s and '30s. The disrepute of the space and experience of coin-in-the-slot machines was reinforced in this period as new associations emerged with gambling and crime. In a way this was a product of arcades substituting one kind of amusement for another, with recordings of music and dramatic acts phonographically and cinematically being succeeded by a variety of games. Along with the shift from watching or listening to playing, the emporium's name sometimes changed. What had been an arcade or an automatic vaudeville was now sometimes a playland or sportland, though still retaining the same old characteristics of being a space for coin-op play open to pedestrian traffic and free to enter.[21] (The name "arcade" also persisted, of course.) Coin machines continued to be installed in numerous public places, as they had been for decades, including cigar and candy stores, railway stations and bus depots, hotel lobbies, gas stations, barber shops, and lunch rooms.

The new game emporiums contained many attractions, from diggers and cranes to love testers and mindreaders.[22] Many coin-op games of the time were versions of popular sports such as horse racing and baseball, and shooting games were an attraction whose appeal spanned virtually the whole century, as electro-mechanical rifle shooters eventually gave way to *Space Invaders*, *Asteroids*, and *Defender*. But the most popular type of coin-op game of the Depression years and decades to follow was the pin game, and until video games surged in popularity at the end of the 1970s it was the pin game that epitomized the trade in coin-operated play.

The rise of pinball during the 1930s has often been associated with the hardships of the Depression. Down-on-their-luck players found solace and amusement in their few moments of diversion, and the very low cost made it affordable even to those unable to enjoy any other commercial leisure. "It only took a few pennies in a *Baffle Ball* or *Ballyhoo* machine to wipe away the gloom of unemployment," according to one popular history.[23] The pin game was also seen as a boon to proprietors of candy and drug stores, who kept countertop games by the register, hoping that customers would drop some of their change in the slot. This vision of the pin game as a solace in hard times was counterbalanced by more fearful and puritanical rhetoric of social reformers, which in some ways was rehearsed again in the 1980s when pinball gave way to video games as the public amusements most fancied by young people. There is one big difference, however, between antipin and antivideo crusades: pinball was more often and more strongly identified with gambling and crime, and its bad reputation was a product not only of cheapness but also of vice. The low reputation of public amusements suffered from this identification for most of the twentieth century.

Pin games have a long history, and in different moments they have been games of skill and chance, games paying out rewards in cash or other prizes and games "for amusement only." Like many forms of play, a pin game is something gamblers can bet on, and a great many surely did bet on the outcome of a game of pinball. The history of pinball is also one of technological innovation and adaptation to new environments of regulation and cultural acceptability. The electronic game with a steel ball, a plunger, a playfield and backglass, bumper-thumpers, electrical noises, and two flippers was actually relatively late in pinball's history to emerge. What was called pinball in the days of the Depression was a rather different object.

A pin game was any game on which one or often more balls or marbles rolled along a board pocked with holes or studded with pins for the ball to strike, altering its course. One key game of the 1930s, *Little Whirl Wind*, was upright like pachinko, and a penny dropped in its slot bought the player five balls to be moved through a spiral maze and down to holes below. Scoring would depend on which holes the ball fell into. Hitting or shaking the machine helped the player. Some pin games were modeled on popular sports. Gottlieb's *Baffle Ball*, one of the hit amusements of the 1930s, was a countertop version of baseball with pins surrounding holes marking four bases in a diamond. Some games, like Bally's *Rocket*, paid out cash like a slot machine to winners who achieved a high score or struck a target. Payout games were particularly popular in the later 1930s.[24] The decades to follow saw tension within the coin-op business between payout pinball and games for amusement only. Bally, a company named for its hit *Ballyhoo* pin game of 1931, was mainly a payout game manufacturer. It introduced its lucrative *Bingo* pin game in 1951. The *Bingo* player tried to line up numbers in the backglass by making balls fall into numbered holes in the playfield, and Bally proceeded to make bingo games of various themes and versions. Scoring in a bingo made money for the player but also courted trouble with the law. A federal court in 1956 distinguished between games of chance, like bingo, and flipper games, which were not regulated as gambling machines.[25] Dodging trouble, the coin-op industry preferred flipper games in the wake of this decision and often substituted the name flipper for pinball, to avoid any lingering negative connotations.

Long before this setback for the amusement industry, pin games had courted serious trouble. However factual or imagined, an association between pin games and crime rackets developed during the 1930s and endured into the video age. Public opinion of pinball was typically that of "an evil that threatened the morals of the nation."[26] Along with disapproval of gambling, this prompted municipalities to ban pin games as they did in New York, Chicago, and many other cities, just like one-armed bandit slot machines. A famous image of New York City's Mayor LaGuardia destroying coin-slot machines in 1934 crystallized the prohibitionist fervor of the age. It was also familiar from popular media, such as the Warner Bros. picture *Bullets or Ballots* (1936), in which gangsters force a merchant to place nickel pinball machines in his store across from a primary school, profiting from the pupils' misspent lunch money. As *Better Homes and Gardens* put it in 1957, pinball machines "can wreck the civic enterprise and economic well-being of any village, town, or city."[27]

At the time of pinball's greatest popularity and profitability in the 1970s, expectations of its potential harm to children persisted even if the threats had shifted with the times from gambling to sex and drugs. Popular press articles on public amusements would reference the questionable reputation of arcades, where children might be "easy victims for sex criminals, narcotics peddlers, and others." Any indication of rising fortunes for the coin amusement trade or the newfound respectability of pinball would also reference its past in seedy dives and its underworld connections as "a slot machine in disguise."[28] Residual mafia associations persisted as well, as reflected in a 1977 episode of the CBS television series *All in the Family*.[29] When Archie Bunker buys Kelsey's tavern and goes into business as a saloonkeeper, his wife Edith attempts to make over the establishment with a feminine touch of taste and class. She wants to bring in new tablecloths and replace photos of prizefighters behind the bar with portraits of George and Martha Washington. "We'll get rid of the pinball machine," she says, "'cause that's gambling." Archie, exasperated, answers: "You get rid of the pinball machine and a guy who smells like garlic will come around, kiss me on both cheeks, and put a hole in my head!"

The surge in pinball's popularity in the 1970s provides a context for this moment of *All in the Family*, when pinball was very familiar and yet still sometimes felt to be a problem. Its popularity was caused by a cluster of factors, including the introduction of solid-state electronic pinball machines and the repeal of pinball bans in cities such as New York and Chicago. But not least of these factors was the centrality of pinball to the narrative and imagery of the Who's rock opera, *Tommy. Tommy* was released as a double-LP album in 1969, and adapted into a feature film released in 1975. The track "Pinball Wizard" climbed the pop charts twice, first for the Who in 1969, and then for Elton John, who plays the Wizard in the movie, in 1976. In the sequence in which he appears, John's outlandish character competes against the eponymous hero in a televised pinball competition staged spectacularly as a musical number backed by the Who onstage in a theater filled with spectators, making hard rock music and pinball competition into companion popular diversions. Pinball had originally not been central to Pete Townsend's conception of the story of the opera's "deaf, dumb, and blind" character, but he apparently added the game as a rebellious kind of amusement to fit with the musical style of the work (and more generally, of the Who). The popularity of the music and movie promoted pinball among young music-loving amusement hall denizens, and Elton John appeared in advertising to promote Bally's tie-in pinball machine, *Capt. Fantastic*, which played on his characterization of the Pinball Wizard in the movie. Another spin-off from the movie, a pinball machine called *Wizard*, was also popular in the later 1970s. One coin-op industry observer commented that his trade "can thank the rock and rollers for lending an aura of hip legitimacy to flipping the silver ball."[30] The *Tommy*-themed pinball machines were part of a wider development of cross-promoting entertainment media with coin-op amusements, in particular to appeal to young, white, male players. Other examples of these 1970s tie-ins were pinball machines with pop culture or sports themes such as *Superman, Close Encounters, Buck Rogers, Star Trek, Dolly* (Parton), *KISS, Evel Knievel, Charlie's Angels, Bobby Orr,* and *Playboy.*

Some of these machines such as the *Playboy* model were also new in a different way: they were marketed to consumers as objects to purchase for the home. Coin-op games had been made for use in public until this time, and the public amusements trade was surprised by the interest in pinball ownership by individuals. This might have been partly a response to pinball bans: if they were unavailable to play in public, one option for fans of the game was to acquire them for play in private.[31] It was also partly a product of pinball's rising legitimacy and its growing popularity. Pinball was in a moment of rescue from the bad reputation of the past, partly out of the same Eisenhower-era nostalgia that produced *Happy Days* and *Grease,* but also as a product of a new context, years removed from the association of public amusement with gambling and criminal gangs. Pinball enjoyed the respect of intellectuals, and a number of popular illustrated histories were published in these years reclaiming pinball as an object of appreciation alongside the cinema and other popular arts.[32] The home market was perceived to be an untapped growth area, and the acquisition in 1976 of one of the major pinball manufacturers, Gottlieb, by a Hollywood studio, Columbia Pictures, was a response to this opportunity for expanding the home market.[33] The acquisition of Atari by Warner Communication, Inc. (WCI), in the same year was similarly an opportunity to capitalize on play in the home as well as in public places (Atari made both pinball and video games for the coin-op trade) and a sign of the rising legitimacy and profitability of public amusement. It also marked games' emergence into the realm of media where they might be a catalyst for the synergistic marketing of cross-promoted franchises such as *Superman,* which was a comic book, paperback book, movie, and video game all under the WCI umbrella.[34] The presence of video games in both arcade and domestic space suggested a similar potential for pinball to expand into a consumer market in places where the game might earlier have been shunned for its underworld stigma.

Still, these developments drew on the rebellious nature of the pinball player, the romantic outlaw character in the style of James Dean.[35] Pinball, like rock and roll, connoted a vaguely threatening low culture likely to arouse the disapproval of one's parents and other guardians of civilization. It was a fixture of tough places like saloons and urban arcades supposedly filled with lowlifes. But it was also a rather harmless game, a source of simple pleasures. These youth culture connotations, and a lingering association with gangs and hoodlums—with tough guys in leather jackets, cigarettes dangling from their lips—made pinball seem sexy. So did the racy backglass art showing off the curvaceous female form on many machines of the 1970s alongside motorcycles, spaceships, and other boy-fantasy scenery. The tough, competitive culture of pinball and the sexualized backglass art alike were aspects of the masculine character of 1970s arcade amusement. And the rising reputation of this culture was a function not only of shedding past negative associations, but also reproducing them as a kind of safe danger, a contained risk. The tough, working-class, youthful masculinity conveyed by the image of the pinball player in popular imagination was part of the romantic appeal of the game.

The Game Room in the Age of Pinball

Bruce Springsteen's "4th of July, Asbury Park (Sandy)," released in 1973 on side one of *The Wild, the Innocent, and the E Street Shuffle*, romanticizes a beachside resort in New Jersey: a boardwalk and pier out over the Atlantic Ocean, midway rides like a Tilt-a-Whirl, a fortune teller, young people out late, holiday fireworks. Local "wizards" play in a "dusty arcade" filled with "pleasure machines." The people and places, as in so many of Springsteen's lyrics, are from the lower end of the class hierarchy: factory workers and waitresses, pale-skinned greasers in high-heeled shoes with shirts open, "cheap little seaside bars," cops making a bust. The pinball room is for youthful carnival diversion. Young men like the first-person voice of the lyric amuse themselves on the cheap and compete with one another before moving on to drinking and sex. The pleasure machines in the dusty arcade are places for working-class, maybe "ethnic" white youth, teenagers, or men in their twenties out for a good time in a town whose economy is driven by commercial leisure.

There may have been electronic pleasure machines like *Pong* in the dusty arcades of Asbury Park by 1973, but Springsteen is pointing to an earlier period in electro-mechanical coin-op play. Amusement park game rooms of the 1960s and '70s would have typically combined a number of different kinds of attractions: shooting, driving, and sports games; old-time novelties like strength and love testers; shuffle and bowling games; and more pinball than anything else. They might have also included a photo booth or some kiddie rides. Resort town arcades often would include redemption games that reward good play with tickets to be collected and exchanged for a prize. Skee ball had been popular for decades. In earlier days such machines could be operated with a penny, nickel, or dime, but by the 1970s the coin was a quarter, which would buy you five balls of play on a flipper. Perhaps because of their association with youth and with times marked off for leisure—evenings, weekends, and holidays—these pleasure machines have often been treated, as they are in "Sandy," in nostalgic terms as tokens of moments past and experiences of innocence or the trials of coming of age.

Another popular culture representation from less than a decade later shows a rather difference space and a different image of leisure and play. In a scene early in *Fast Times at Ridgemont High* (1982), the camera pans across a number of video-game screens and teenage boys playing the uprights. This is part of a longer sequence establishing the southern California shopping center as the central site of high school students' socializing. The kids work at the mall and play at the mall. After establishing the arcade space, one character approaches another about buying rock concert tickets, and this commerce of questionable legality occurs against the backdrop of a *Space Invaders* and a *Pac-Man*. Nearby the arcade is a movie theater and a counter selling fast food, and just outside its entrance are apparel shops like the Gap. The scene ends as the ticket broker, Damone, puts an arm around his more innocent friend Brad and says, "Come on, let's find you a girl."

The suburban teenage life occurring here, by contrast to Springsteen's lyric, is much more middle class. The mall is a clean, upscale destination for consumer culture and communal gathering. Although ticket scalping is a less legitimate part-time job than waiting tables, it is more respectable than pushing drugs or shaking down shopkeepers for protection money. The two spaces still share an association with young people's romantic desires and conquests, but in other ways, distinctions separate them not just by type of amusement but also by the identities of their clientele.

The video arcade of the 1970s and '80s is much more often associated with the type of representation in *Fast Times* than in "Sandy." It may be tempting to read this difference as one between earlier and later coin-operated amusements. But the shift from working class to middle class, from dusty to sanitized, occurred before video games replaced pinball machines as the central appeal of these play rooms. This shift was the product of a suburbanization of leisure, and an acceptance of previously disreputable coin-operated amusements as a legitimate attraction within the regional shopping centers outside of city centers (as well as in other suburban and exurban locations). In the 1960s and '70s, such places were becoming not only lively retail hubs for middle-class American consumer spending, but also significant social magnets representing a new version of culture and community. The spread of pleasure machines from urban and resort locations to suburban shopping malls came with a rearticulation of space, attaching new associations and meanings to the amusement room.

In renewing and reviving the amusement center in the era of suburbanization, retailers and operators sought to cleanse coin-operated games of their lingering disrepute. The seedy image of the urban or seaside game room was an impediment to the arcade's acceptance by shopping center landlords, and the coin-operated trade made efforts to shed this image. The old idea of a game room was a dilapidated downtown or oceanfront penny arcade home to a cast of shady characters: sailors, hobos, undesirables. The new idea was to make the amusement arcade "first class" and to ensure "wholesome fun." Rather than urban arcades or sportlands, catering in particular to young men, the shopping center amusement rooms might have more family appeal. They could be to the 1970s and '80s what the corner drugstore, with its soda fountain counter, was to the 1940s and '50s.[36]

While many of the suburban arcades were in free-standing locations or amusement centers including other diversions like miniature golf, go-karts, or batting cages, the shopping mall arcade was particularly commonplace and reliably profitable. Some chains, including Aladdin's Castle, ran arcades in both kinds of locations. The game room attracted teens to the mall as a meeting place and hangout, but also gave youngsters a way to pass the time while their mothers shopped.[37] Teenage mall visitors in the 1980s might have gone there for the video games, but boys also went to pick up girls and vice versa.[38] Male players far outnumbered female ones, often by a ratio of four or more to one. Many of the social pleasures of the arcade were similar to those of earlier generations of gendered amusement, and more likely homosocial than romantic: competition, challenge, proving oneself to peers.[39]

The mall was not the site of the majority of video arcades in the 1970s. In 1976, shopping centers accounted for only about one-third of amusement rooms.[40] The remainder were in street spots, resorts, or amusement or theme parks like Golf N Stuff (GNS). But the mall was a fast-growing location for coin-op play, and the mall arcade concept led the way in reconceiving the game room as a safe and respectable space for teenagers to congregate, putting to rest the old stigma of pinball and other penny arcade fixtures. Old arcades such as those in New York City's seedier districts like Times Square might have "gone to the dogs," but new, suburban spaces were considered surprisingly nice by old standards.[41] Some chains of amusement parks including GNS and Aladdin's Castle also ran shopping center locations, and by the end of the decade the independent retailer hoping to open a mall arcade would have found a market effectively closed to new competition and dominated by chains.[42] The prejudices of mall landlords had been overcome by this new concept of the game room, along with the evidence of profits to be shared by the retailer.[43]

The new arcade followed in many ways from the configuration of suburban shopping center space, from its meanings and functions. Shopping center arcades developed an identity from their surroundings, and could only be acceptable to landlords and other tenants, and presumably to many shoppers, by fitting in and conforming to the sensibility and purpose of the location. The shopping mall was developed not only to be a regional commercial venture to exploit the consumption desires of suburban residents, but also to function as a nexus for community life in the newly populous regions on the outskirts of cities. The architect Victor Gruen, often credited as one of the pioneers of the mall, argued that it would not be sufficient merely to sell consumer goods to suburbanite shoppers. The success of the shopping center would be realized by meeting further, deeper needs: making "opportunities for social life and recreation in a protected pedestrian environment." In this way, malls would be a place for "modern community life that the ancient Greek Agora, the Medieval Market Place and our own Town Squares provided in the past." The shopping center was to marry many functions in one spot. Its role in sustaining "civic, cultural, and social community" life would be met not only through retail shopping but also by a diverse array of other offerings to the suburban public.[44] These included artistic and cultural displays and performances, educational and recreational programming, and spaces conducive to meetings and gatherings. Malls in the 1950s, '60s, '70s, and '80s contained eye-catching fountains and sculptures in central courts or atria. They staged fashion shows, art exhibitions, concerts, and dances. Many offered rides and ice-skating rinks for amusement and recreation. Play areas for young children were also common, with slides and climbers, and perhaps rides. The design of a galleria, with two or more levels overlooking a central concourse—or perhaps a clock tower, fountain, skating rink, or fifteen-foot-square chessboard—made for social gathering places and communal spectacles of public life. A mall, one architect wrote, should "relax and refresh the families who use it, and promote friendly contact among the people of the community."[45]

In contrast to the city streets and urban shopping districts, the suburban shopping center had the distinction of being predictable and comfortable.[46] It was a new kind of public area cleansed of the less desirable elements of city life, such as poverty and filth, sheltering its visitors from becoming troubled by unpleasant things. One strong appeal of the mall was its enclosure, its protective function to control the environment and keep out undesirable imperfections of social reality.[47] The very qualities of public amusement spaces that had led to their disrepute—their associations with marginal persons, with crime and dereliction, with seedy or dusty or shady environments—were absent by design from suburban shopping malls. Also often absent were racial and ethnic minorities, some of whom brought legal actions claiming discrimination in leasing or hiring, or being harassed by security personnel.[48] The shopping center's enclosure and protection were to facilitate a particular ideal of community in the suburbs that would reinvent the bonds of small town life, the village green or main street where neighbors and citizens would meet. But this was a class and race fantasy of protection from the other, insulation from the trouble of a racially and economically segregated and stratified society.[49] Commentators on shopping malls have often compared them to Disneyland, which speaks to their function of giving visitors pleasure in a nice public environment, but also to their ersatz nature as a place that merely seems like an authentic town center, encouraging a longing for simpler, happier times. Like a theme park and an upscale mall, an amusement arcade depended for its legitimacy and respectability on associations between leisure-time pursuits and social identities. The safety of a white middle-class or affluent clientele helped the arcades of the 1970s dodge the stigma of public amusements and renew the image of pinball and other coin-op games in popular imagination. Now they were not merely game rooms; they were "family fun centers."[50]

Proprietors of these centers adopted a number of strategies to secure this revision in their trade's reputation, but the most central, most frequently mentioned in discussions of coin-op games and respectability was cleanliness.[51] A nice arcade, no matter the location, one that would avoid trouble and keep quarters dropping into the cash box, would be first of all a clean arcade. A clean arcade, with well-functioning machines, supervision, and rules prohibiting unwanted behavior, would be unlikely to cause problems for the mall and its patrons. It would be deemed worthwhile by the suburban communities whose civic life was now to be found in a shopping center, which was itself a pleasant model of cleanliness. It would be considered safe for the children of the suburban bourgeoisie.

Cleanliness in these discussions was both a literal and figurative term. It meant sweeping, dusting, and picking up trash, but it also meant casting off the connotations of the "dusty arcade" of Bruce Springsteen's memories. A clean arcade would be safe from lowlifes and beggars, small-time criminals and hustlers, adults who think they're still teenagers, tough guys sneering at genteel sensibilities.[52] It would fit into suburbia and its American middle-class culture in an era of "white flight" from the city. It would put distance between the new arcade and its other, the old amusement center frequented by the wrong kind of crowd. Clean also meant legal, free of corruption and vice, a legitimate business. Clean even suggested virtue. "Good clean fun" isn't a reference to an absence of dirt; it affirms a moral judgment, a sense of wholesomeness and propriety.

In the coin-operated amusements trade, which included not only pinball and video game machines but also other arcade amusements, pool tables, and jukeboxes, the push for respectability through cleanliness and orderliness was pervasive during the 1970s and '80s. This was good business logic, assuring the trade of sustained and increasing revenue through not only the acceptance of pinball and video games by polite society, but also increasing freedom from onerous regulation. If coin-op arcades were good clean fun, there would be no cause to ban them or the machines within their walls.

Cleanliness was not a new ambition in the 1970s, having been urged on arcade owners as early as a 1906 item advocating for "cleanliness and light" in penny vaudeville parlors.[53] A 1933 *Billboard* story about the rise of pin games encouraged keeping arcades well lighted and clean as signs of refinement, the better to overcome "the feminine prejudice."[54] In the 1970s, though, cleanliness was linked not so much with light and refinement and gentility as with orderliness and respect for authority and property. It was often mentioned in the same breath as maintenance. If machines were kept running in good working order and the floors were kept free of dust and trash, the arcade would appear respectable.[55]

Enforcing a high standard of cleanliness went hand-in-hand with forbidding behaviors such as smoking, eating, and drinking. Signage declaring these prohibitions would function not only as ways to prevent trash and ash from collecting on the floor or dirtying the machines, but to maintain a sense of respect and discipline among the patrons of the arcade through their obedience to posted rules.[56] In upscale arcades such as the Time Zone chain, which opened in 1974 and expanded in southern California during these years, a "code of conduct consistent with intelligent maturity" was enforced as a way toward achieving "profits, prestige, and permanence in the leisure industry."[57] Trade wisdom was that rules must be enforced by arcade managers, and troublemakers kicked out. "Dirt and garbage lead to more dirt and garbage, but clean tends to stay that way," according to one trade journal editorial. It urged a high standard of policing the arcade against these perceived threats, measures undertaken though the discipline of authority and a respect for orderly conduct. "There are many people in this country," it warned, "who honestly believe that arcades are settings for everything from drug peddling to prostitution. The arcades will stand guilty until proven clean."[58]

To be proven clean, arcades posted their codes of conduct on the wall for all to see. The terms might vary from place to place, but certain themes were consistent. In addition to forbidding food and drink as a way of keeping the space free of litter and mess, banning drinking also regulated consumption of alcohol. Likewise, no smoking meant that cannabis products would be proscribed along with tobacco (many arcades still had cigarette machines, though). Posted policies regulated the age of patrons, making the arcade a safe space for children. Kids were unwelcome during school hours and at night they might need an adult chaperone. The number one rule, according to a survey of operators of coin-op machines in 1978, was "no horseplay."[59] This ensured that the environment of the arcade would be peaceful and respectful, like the mall beyond its doors. Young people spending their money in an arcade would follow the same expectations of polite and respectful conduct as in school or church—or so the coin-operated amusements trade hoped.

Arcades in many locations other than malls, from bowling alleys to college unions, often had their own site-specific expectations. Some of the new arcade chains, such as Time Zone, Aladdin's, and GNS, were regarded as "plush" family fun centers, a reference not only to their carpeting but also their luxe appeals to an affluent family trade.[60] A coin-op trade paper celebration of "Arcades Today" depicted them as transformative for the business. "The metamorphosis of arcades from the center city sleaze or earlier times to the plush carpet centers of today has been heartwarming."[61] Plush carpeting would have been hard to find in dusty arcades and sportlands of urban and resort areas. So would these new arcades' elaborate, Disneyesque décor featuring themes such as medieval castle or western ranch. The clientele in such locations were "All-American folks from central casting," which meant they would cause no trouble and enhance the emporium's classy image. These plush amusement centers located in miniature golf courses and shopping malls emerged first in southern California, appearing to be "not unlike something you'd see concocted in the special effects departments of their own Hollywood studios."[62] The businesses invested unusual sums in advertising and promotion, were kept scrupulously clean, and featured an unusually high percentage of TV games, which carried less residual shame than pinball.

During these years, Ramada opened arcade rooms in its chain of inns to capitalize on the popularity of coin-operated amusements. Like a suburban mall, a nice hotel would want to ensure that pinball and video game machines would be integrated into its space without bringing along undesirable associations. Ramada managed this by stressing the cleanliness of its game rooms, eliminating any hint of a "carnival look" or any suggestion of a "garish, bus station appearance." So that dust or trash would have no place to gather, Ramada had the legs removed from pinball tables and mounted them instead on carpeted benches. Rather than placing the machines within earshot of a lobby area, the arcade was located behind Plexiglas to eliminate the noises of electronic machines. The mall arcades catered to young people in particular, but both pinball and video games were popular with players of many ages. As a place of lodging for business travelers, Ramada appealed to adults as well as children with its game rooms. Almost two-thirds of the quarters spent in its arcades came from the pockets of businessmen. Like the rules posted in mall arcades, the efforts to remove any hint of stain or stigma from amusement machines in hotels upgraded the class status of coin-slot play and made it newly respectable.[63]

This respectability arose from a number of sources, and the inclusion of video uprights within the game room was one of them. But at the beginning of 1979, when the story of Ramada's game rooms was reported, pinball was still the most popular and profitable coin-op machine in the arcade. The shifting status of public amusements happened during pinball's reign, and video games surpassed pinball only once change was well underway. The emergence of clean, plush, family fun centers occurred along with the emergence of video games, but at first these new arcades had a much more varied array of offerings than they would by the early 1980s, when the new electronic games now reigned over all others.

From Pin Games to TV Games

By the early 1970s, the coin-operated amusements business had for years been focused most of all on jukeboxes and the establishments in which they could be found: those, like *All in the Family*'s Kelsey's, serving mainly alcoholic drinks. A jukebox would require regular servicing on the operator's route at least to replace old records with new ones, keeping up with the record industry's cycle of hits climbing the charts. Operators also handled amusement games, some of which, like Kelsey's pinball machine— alongside pool, shuffle, and soccer (foosball) tables—were found in the same bars. In the language of operators, these were "street" locations as opposed to "arcades," which were places defined by their amusement games (though they also often had coin-operated jukeboxes, photo booths, and vending machines). During the 1970s, a number of changes shook up the business of coin-op leisure. First video games were introduced, then pinball became newly popular and displaced jukeboxes as the trade's biggest earner, and finally video games became the most popular and profitable fixture of locations in the arcade and on the street. These shifts caused the trade organization representing the coin-op business, the MOA (Music Operators Association), to change its name in 1976 to AMOA (Amusement & Music Operators Association).

Video games became such fixtures by riding pinball's coattails, and by inheriting some of pinball's cultural status even as they cultivated their own. Video games trailed pinball in earnings and interest for most of the decade, despite the attention they attracted for being new and high-tech. Like 45s in a jukebox, video uprights and pinball machines had to be changed from time to time to refresh the interest of regular patrons. Video games in particular had a strong initial novelty value, and after a period of weeks or months would have to be rotated to a new location where the players might be unfamiliar with them. Success in the coin-op business meant moving machines from place to place on the operator's route to maximize the earnings in the cash box.

Before 1979, video games did not generally become hits. One exception in addition to *Pong* was the unusual Atari product *Indy 800*, a racecar driving game played by up to eight players standing at wheels around a common up-facing display of a racetrack. Each of the players controlled one car on the track, and it cost a quarter for each of the players to race. *Indy 800* was a big and lasting earner for Atari and amusements operators, often occupying a spot in the center of the game room (it could not be placed against a wall like most pinball and video cabinets); but the typical TV game was less lucrative than the typical pinball machine. Pinball machines were also unlikely to become hits to the extent that one would be significantly more profitable than others, or that arcades would want to have more than one of a particular pinball machine.

In the second half of the 1970s, pinball became the most important type of coin-operated machine in arcades and street locations. In annual summaries in the coin-op trade press, pinball was named machine of the year, as in 1977 and 1978, both referenced as a "year of pinball."[64] Advice to the operator and proprietor was to showcase the pinball machines: the "number one attraction in any game room should be a beautiful array of flipper games."[65] When surveyed in 1977, operators said that if they could deal in only one kind of machine, they would choose pinball, followed by pool, jukeboxes, soccer tables, and only then video games.[66] The "fair-haired darling of the coin operator," a pinball machine was a steady if not spectacular source of income.[67] Especially in street locations but also in arcades, customers spent more money playing pinball than they spent playing any other amusement machines including TV games.

The surge of pinball past jukeboxes was a product of some changes already described: the rock-and-roll aura lent to pinball by *Tommy*, the legalization of flippers in big cities such as New York (1976) and Chicago (1977), the legitimacy won by clean and plush suburban arcades, and the effort to market pinball to consumers. The solid-state electronic pinball machine was a technological innovation that renewed interest in the game. Pinball had undergone many technological innovations over the years, and many became standard features of the game, including flippers and bumper thumpers. Replacing electro-mechanical technology with solid-state electronics fit into the wider developments in high-tech in the 1970s, as electronics penetrated many areas of everyday life from supermarket checkout counters to wristwatches. The new solid-state pinball machines introduced beginning in 1976 had electronic digital scorekeeping and electronic sounds, and an increased complexity of play. The mechanisms controlling the game were no longer the old switches and relays but now integrated circuits. This made them more similar to video games, which of course were also a form of integrated circuit electronics. Some players might not have recognized the change from electro-mechanical to solid state, and some games were released in two versions. But to some aficionados the new games would have been noticeably technologically upgraded.

The difference between electro-mechanical and electronic technology also figured in, however, to the widespread disdain for TV games among AMOA members. In surveys of operators during pinball's heyday of the later 1970s, many expressed frustration with video games and a preference for many other kinds of coin-op machines. "Make less TV games" was their refrain.[68] One issue with video games was their unreliability.

Electro-mechanical pinball and other arcade amusements could be more easily fixed by handy proprietors or operators who knew their way around relays and switches. A totally electronic game could only be fixed by someone with a less common expertise in electronics. Since 79 percent of operators found video games to be unreliable, the preference for pinball above other amusement machines made good sense, especially considering that few video games had yet to become very popular and thus seriously profitable.[69] In 1978, *RePlay* asked operators where TV games fit on their route, prompting operators to lament that they needed to be moved and repaired frequently.[70] That the question even needed to be asked indicates video games' questionable status in the era of pinball's reign.

Video games, however, were also crucial to the rising respectability of arcades and coin-operated amusements including pinball even before they outperformed all other amusements. They brought respectability by being newer and more high-tech than flipper games, and they gave coin-op amusements an "entrée," according to one member of the AMOA, into "thousands of locations where any kind of coin operated amusement game was taboo, unacceptable, and not permitted to operate in the past."[71] These places included cocktail lounges, hotels, airports, and malls. Video games had no old reputation as playthings of gamblers, criminals, and hoodlums. They were considered more "intellectual" amusements, which was an indication of their class status in relation to other games. They were, in the tradesperson's terms, "sophisticated, adult, scientific marvels."[72] Video games were also in many middle-class homes already by the later 1970s, which helped public amusements—some of which were now hardly different from home versions made by the same companies—overcome their original sin of association with crime and gambling. The advertisements for video games aimed at consumers were an "image-producing boost" giving coin-operated leisure a "respectability shield."[73]

Another benefit of video games for the coin-op business was a generational development: kids who grew up with video games seemed to like them more than adults who had not, which would serve to renew and refresh interest in amusements. The "family trade" at the plush arcades and bowling alleys liked TV games more than the older crowds at taverns and pool halls. One of the keys to success of a family fun center arcade like Time Zone was "a high percentage of TV games" along with the rest of the trappings of upscale leisure.[74] The college crowd also preferred video games more than other groups. In an arcade that opened in 1977 in West-wood, California, near the UCLA campus, four out of five machines were video games at a time when most game rooms were dominated by pinball.[75]

Late in the 1970s, then, video games were popular in particular with younger and more affluent players of public amusements, and their presence had been essential to the revitalization of public amusements as a legitimate leisure-time activity. But they were not yet impressively profitable for the coin-op trade by comparison to other types of machine. This changed beginning in 1979, when *Space Invaders* became the industry's first enormous hit. By April of that year, *Space Invaders* was the "world's hottest game," creating a kind of demand never seen before.[76] Like many a surprisingly popular craze before it, *Space Invaders* inspired songs such as one released as a 45 RPM single (first in Japan, then imported to the USA) combining the bleepy alien invasion sounds of the game with a disco tune backed by an orchestra, and another with the lyric "He's hooked, he's hooked, his brain is cooked." *Space Invaders* was so hot that a *Space Invaders* pinball machine was made to exploit its popularity. This demand hurt other games in the market when arcade patrons wanted to play only *Space Invaders*, but the surge in interest in video games lifted the entire industry. Arcades that before had never kept more than one of a video or pinball machine now needed multiple *Space Invaders* cabinets to keep up with the thirst to play it. It might not be unusual for one location to have four *Space Invaders*, and American operators learned of an arcade in London's Piccadilly Circus where ten were arrayed side by side.[77] Operators now enjoying increased revenue from *Space Invaders* became frustrated that other games could not match its earnings, and companies tried to duplicate its success with spinoffs and copycats, some legitimate (*Space Invaders Deluxe*) and some knockoffs of the "space creature assault" concept. Video game manufacturers followed up on *Space Invaders* with many more space shooter games: *Asteroids* and *Asteroids Deluxe*, *Missile Command*, and *Galaxian* all collected a great many quarters in the cash box.

Adults as well as children took to *Space Invaders*. A tavern that had one video game and two pinball machines in the later 1970s might have one pinball and two video games after *Space Invaders*. For all players, the novelty status of video games had been shed, and machines rewarded much more devoted play. It became unnecessary to rotate games around the operator's route when there was such sustained interest in individual machines.[78] It might be unfair to say electronic games simply got better, but they newly rewarded devoted, repeated play and attracted players to individual cabinets more strongly. The games had also become more reliable, and complaints about maintenance largely ceased. Even if these issues persisted, the operators were making too much money to disdain video games. By late 1980, a "year of video" in the amusement trade, the average TV game upright had doubled its earnings from a few years before.[79] Now when asked if they could offer only one kind of game, the operators said "video," which was now regarded as the "Rolls Royce" of amusement machines.[80] By 1982, the top-earning video game was collecting an average of $255 a week (the top pinball earner took $157).[81] This was several times the take of a few years before.

At first, a new video game would offer a shorter duration of play for one quarter than a pinball machine. Games were more challenging initially to the unfamiliar or unseasoned player. This brevity of play helped increase the profitability of video games, and is one reason why some pinball machines dropped their play from five to three balls for a quarter at the same time that video games got hot. But unlike pinball, video games like *Space Invaders, Asteroids, Missile Command*, and later *Pac-Man* and *Donkey Kong* and many less popular examples, encouraged players to invest considerable time, energy, and money in increasing their skill and mastering the game. Pinball seldom inspired such devotion. Video games, unlike pinball, would adapt to your skill, becoming harder as you got better. They also had less chance built into them than pinball. A proficient video game player could milk a single quarter for a long duration, spending hours with these "infectious inventions."[82]

Unlike pinball, video games also remembered the machine's high scores, and some allowed players to enter initials next to theirs for bragging rights.[83] Twin Galaxies Arcade in Ottumwa, Iowa, maintained an international scoreboard of high scores on various arcade games.[84] One trade paper headline read, "They Play for Hours, and Hours, and Hours to Beat the High Score!" In the early 1980s, many news reports told of young video game players who set records not just of high scores but also of duration of play for various popular arcade games. They would master the game to the point that they could play them continuously until their bodies could endure no more pushing a joystick, pressing a button, and coordinating hand and eye. A fourteen-year-old in Maine played *Asteroids* for 29 hours 35 minutes, and the record for *Missile Command* was 28 hours by a player in Spokane. Atari claimed its record length of play for *Asteroids* was 50 hours.[85] These games were hardly interchangeable, though, and becoming proficient at one did not necessarily mean mastery of others. In contrast, pinball players transferred their skill more from game to game. A pinball machine's identity might come as much from its theme represented in its name, and its backglass art and playfield design. Young video game players told *RePlay* in 1981 that by comparison to video games, pinball machines are "more similar to one another" and more dependent on luck.[86] Video games were regarded as more reliant on skill. This is partly why so many photos of video arcades from this period show observers looking on at someone else playing. They are there to watch and learn, to gain insight into successful play, to figure out the ways of the machine and the strategies of the best video kids. Many guidebooks were also published to instruct the would-be high-score achiever.

This was a competitive, youthful, masculine culture familiar from experiences of sports and other kinds of contests based on achievements of skill. In representations of arcades in popular culture and the popular press, the player engrossed in a video game, engaged in an effort to master it and achieve a high score, is most often young and male. In television commercials for home versions of arcade games, for example, the player is invariably a teenage or young adult male whose entire being is consumed in electronic play. The image on the cover of *Time* magazine from January 18, 1982, was of such a young man armed with a futuristic space pistol inside the screen of an arcade cabinet. This obsession with video games, which captured the energy of so many young people in these years, also produced a sense of crisis among adults beginning in 1981. Once video games had become so popular, antigame crusades kicked into gear. Many municipalities that had not reacted against the popularity of pinball in the 1970s sought to regulate or ban video games and video arcades when they became even more popular in the 1980s (a topic to be considered in more detail in chapter 5). The outcry over electronic play was a product of many fears and concerns, but central to its mission was a worry about the force computer games seemed to exert over their players—what Sherry Turkle in her 1984 book on computers (and her commentaries on television and in the press) called their "holding power."[87] Although its effects were often feared, pinball was never regarded as exerting this kind of force over its players. Video games displaced pinball as the central attraction of the arcade in part by offering a stronger, more mesmerizing appeal. But the space of the arcade, and many of its key characteristics and connotations, already existed at the time of the debut of *Space Invaders*. The video arcade and the pinball arcade and the clean and plush family fun center were all part of the same emergent space of amusement for the American suburban bourgeoisie, and its sons in particular.

Figure 1.1
Time magazine's January 18, 1982, cover pictures a young man fighting an alien invasion within the representation of an arcade game.

Home and Away

As a technology and medium used in many locations, the early video game was not defined by one space alone. As later chapters will argue, the home was in many ways as important as the arcade for the establishment of coherent and lasting meanings and associations for video games. But these meanings and associations came into being in relation to a history of public amusements, and the video arcade was heir to a lineage of coin-operated play. The domestic sphere has long been feminized, associated with women, with the feminized labor of childcare and housework. The video games made for the home were marketed as a new way of reintegrating a companionate nuclear family, much as TV had been asked to do upon its emergence a few decades earlier as a mass medium and consumer good. But in order to give video games cachet and to make a powerful appeal to the young males who became their largest group of devoted players, marketing of home games also drew upon the meanings of the public arcade with its excitement and sophistication. The home game was a way to bring the arcade and its noisy, boisterous thrills into the safety of the familiar family space.

The arcade also led the way in establishing the hits that would ultimately sell the home game consoles. *Space Invaders* was not just a success in public play rooms; it also made Atari's VCS into a huge seller, and imitations also helped to sell home computers and rivals to Atari's video game console. After *Space Invaders*, Atari found further success with ports of hit arcade games including *Asteroids*, *Missile Command*, and *Pac-Man*. Every console had versions of popular arcade games, but it was unusual if not unheard of for a popular video game to begin as a home console cartridge and later be ported to an arcade cabinet. The arcade gave a game its legitimacy and established its player appeal with the most central group of video game aficionados: middle-class young men and boys. The home version was a way to play without spending quarters and going out, or without encountering the residual and lingering threats of coin-op play in public. It was good for younger children or anyone who felt unwelcome or uncertain in public amusement rooms.

These threats were integral to the definition of video arcades as new public spaces of play, and they informed the emerging medium's identity in a number of important ways. The video arcade that arose from the history of dime museums, penny arcades, drugstore counter pin games, playlands and sportlands, and family fun centers carried along associations from all of these, some of them contradictory meanings alternately signifying danger and security, corruption and good clean fun. The adult, unruly, cheap, and coarse reputation of pinball and other public amusements remained, but it was counterbalanced by the intellectual, high-tech, and safe suburban appeals of the new forms and spaces of play on offer in the 1970s and '80s. Video games were a new breed of cool recreation, and they stimulated obsessive devotion from young people growing up in the electronic age. Their appeal was the product not only of the technologically advanced pleasure of human minds facing off against machines programmed to challenge and vex them. It also came from the rebellious charge given off by coin-operated amusements spanning most of a century of commercialized, mechanized leisure.

2 "Don't Watch TV Tonight. Play It!" Early Video Games and Television

"Let's play football," one kid suggests. "Tennis!" insists the other. They compromise on hockey. But instead of gearing up and heading for the front door, they troop into the living room and flip on the television set.
—*Newsweek*, 1972[1]

A video game is something you look at on a screen (along with something you hear on synced audio), and also something you play using an electronic interface. In the early years of video games, the video component was most typically a cathode ray tube (CRT) display, more commonly known as a TV set, and video games were a new kind of media device that included TV in its technological ensemble. The display of an arcade cabinet TV game would be used for play only, but the home versions plugged into the same appliance that had been receiving broadcasts over the air. The histories of video games and television overlap, and the emergence of games as a new medium can hardly be understood without considering the televisions that were among their essential components.

Television was at the core of the new technology's identity. Video games were marketed as "tele-games," and coin-op video games were more commonly known as TV games in the 1970s. One early home game console was called "TV Fun." While the meanings of video have shifted over the years, in the 1970s the word was still—as it had been for decades—a synonym for television. It was at the time also developing additional connotations in distinction to television particularly thanks to videotape, video art, and also video games, all of which were uses of television technology for purposes other than watching live network broadcasts.[2] Video games could have come to be known as electronic games, computer games, or something else, but their identity was established in terms of the familiar TV screen and its meanings.

Early games, particularly those available to play in the home, were a new use for a familiar technology, renewing the identity of television and drawing on its cultural status to develop meanings for the new medium. Television and video games became distinct media by the 1980s, but their histories also have something important in common. Video games were understood in the early and mid-1970s by reference to existing media— other electronics devices, other toys and games, and other uses of television sets. Before they were their own thing, video games had an ambiguous status. Video games drew on TV's familiarity to appeal to consumers. But they were also distinguished from television and presented as an improvement on it. They were TV, but more than just TV.

As they are now, electronic games in the 1960s, '70s, and '80s were a heterogeneous assortment of artifacts and experiences. These included mainframe and minicomputer programs in universities; coin-operated amusements often placed alongside pinball machines in bars, bus stations, and shopping centers; handheld devices resembling calculators; and plastic boxes inputting to a television set. Even though they used LEDs (light-emitting diodes) rather than video displays, handheld games of the 1970s such as Mattel *Electronic Football* were precursors of Game Boys and mobile touchscreen devices. Even though they were used by a small number of hackers, early computer games like *Spacewar!* were precursors of PC games and multiplayer online play. Video games in the 1970s and early '80s were also part of wider contexts of play that included Dungeons & Dragons and other nonelectronic games. Video games that plugged into a TV set like the Atari VCS were central to setting an identity for the new medium, but they also overlapped with all of these other phenomena in the emergence of electronic gaming. Arcade games built on the mainframe and minicomputer games. Home TV games adapted arcade games, and handhelds offered similar experiences as well, including driving, tank, pinball, and sports games, and later ports of popular titles such as Coleco's *Pac-Man*, *Donkey Kong*, and *Frogger*. TV games cannot be isolated from the rest, but television technology was essential for the development of these devices. The arcade version of *Pong*, for instance, was described in a magazine in 1974 as "a miniature computer attached to a television screen."[3] Television, not only as technology but as a cultural form of commercial broadcasting, was also essential for the establishment of the new medium's public image.[4] Both the familiarity of TV as an everyday object and its perceived failures as a mass medium were backgrounds against which video games were considered. The new meanings of video games were the product of a history of thinking about media and their social circulation and significance.

Video and the Mass Society

Video games were often imagined not merely as a new use for television but as an improvement on television, turning a disreputable, passive medium into one more active and purposive. This occurred in a wider context of change in electronic media. Television's public image in the period of electronic games' emergence was often one of a social problem in need of solving. Following the FCC Chairman Newton Minow's "Vast Wasteland" speech to the National Association of Broadcasters in 1961, TV was widely considered not so much as a medium of cultural or personal expression as it was a threat to society. The institutional structure of commercial TV networks and stations was regarded as a failure for democracy. Television programs were generally considered to be ephemeral trash. As one famous quip of the period goes: "Television is a medium, so called because it is neither rare nor well done." The experience of TV was characterized by passivity, and the association of TV with the mass audience of consumers whose attention would enrich networks and sponsors tinged this criticism with negative gender and class associations.

The public reception of video games in the 1970s was informed by decades of thought about media and their power to shape modern societies. By the time TV games came along, television was well established as an emblem of media exercising political power and social control. Conventional wisdom about TV at the height of the American three-network era was the product of alarmed midcentury thinking about a cluster of "mass" concepts: mass society, mass media, mass culture, and mass audience.[5] These terms are inextricable from the broad themes of the twentieth century: totalitarianism threatening individuals' liberty, a contest of ideologies (communist, fascist, and liberal democratic), and the Cold War politics of East versus West, which endured until the end of the 1980s. The rise of mass media early in the twentieth century had long been seen as enabling mass propaganda to serve the interests of those in power, influencing the opinions and beliefs of vast populations or even hypnotizing them into subservience to the dominant ideology. The development of broadcast media, first radio and then television, extended the speed, efficiency, reach, and concentration of media. To many detractors of modern media and many observers concerned about their social and political impact, this meant furthering corporate and state dominance over the spread of information and ideas, a means of controlling a mass public under a powerful centralized regime.[6]

In the eyes of many social critics, from Frankfurt School Marxists to American sociologists, from academicians to popular commentators in magazines, mass media had the power to shape a mass society.[7] This could be as true in fascist societies under Mussolini or Hitler, or communist societies under Stalin and his successors, as in advanced capitalist, democratic societies of the West. The citizens in a mass society were not merely subject to the manipulations of official propaganda, but were also believed to have been atomized and estranged from one another, losing their sense of self and community, too eager to submit to bureaucratic, authoritarian leadership. The population was homogenized but also alienated, with traditional community and family ties weakened by urbanization and industrialization.[8]

Media replaced traditional community life for these modern citizens. In the more paranoid versions of this critique, media could be used in brainwashing the masses, compelling their subservience under a totalitarian system through techniques of thought control. A key example in such critiques was the 1938 broadcast of *The War of the Worlds* over the Columbia Broadcasting System radio network, which offered an instance in which mass behavior was supposedly engineered through the power of mass communication. Another hugely influential text dramatizing the power of mass communication was George Orwell's dystopian 1948 novel *1984*, a portrait of totalitarianism in which thought control is practiced through a mass media technology of both propaganda and surveillance, the telescreen.

C. Wright Mills, the maverick sociologist who was an inspiration to the 1960s student protest movement, was one source of mass society and mass media thought that informed the reception of video technology in the 1970s. In his 1956 book *The Power Elite*, Mills addressed mass media as a component of the mass society, which for him is in competition with an alternative social arrangement, a "community of publics."[9] Mills saw modern capitalist society like that of the United States to be governed by a small elite of powerful men who would not necessarily advance the interests of the people they ruled. Democratic societies are premised on the participation of publics in their communities, but a mass society replaces these small-scale conversations with mass media: one voice speaks "impersonally through a network of communication to millions of listeners and viewers."[10] In a public, individuals have the potential to answer back to communications they receive, and "as many people express opinions as receive them."[11] But in a mass society, the flow of information is in one direction only, and most people have no opportunity to respond to mass media, which inhibits them from having thoughts of their own. In a mass society, Mills wrote, "the public is merely the collectivity of individuals each rather passively exposed to the mass media and rather helplessly opened up to the suggestions and manipulations that flow from these media."[12] Mills did not see the move toward a society under the rule of the power elite as simply a product of new technologies of communication, but he credited the mass media for being "the most important of those increased means of power now at the disposal of elites of wealth and power."[13]

By the 1960s, television had become the epitome of mass media: the most popular and profitable form of news and entertainment, dominated by only three national networks, all of them advertising supported. Ideas about mass media as propaganda had been formulated when newspapers and movies were the main forms of commercial media, but radio and television broadcasting were observed to be even more influential and powerful in engineering a mass society. The association between TV, as a domestic appliance, and the great audience of women, children, and people of lesser class status than the elite opinion leaders and corporate and state leaders, led to its identification as feminized and lower-class—mass culture. Mass culture meant a mass-produced and mass-distributed mass media, but also a culture for the mass audience of the mass society. Its reputation was low for multiple reasons: the idea of its audience wanting cultural status, but also its function being the manipulation of that audience, a form of social control. This notion of TV's instrumental role within the mass society was not merely a matter of intellectual debate, but was disseminated in all kinds of popular culture such as Frank Zappa's 1973 song "I'm the Slime" in which he speaks as television content to its audience: "You will obey me while I lead you," he sings. "Your mind is totally controlled." A vernacular critique of mass culture, mass media, and mass society during the 1960s and '70s figured TV as its most perfect weapon, narcotizing and abusing the audience, trivializing politics, and forcing conformity and banality on the viewing public. This can be noted in the nicknames given to television: "the boob tube," "the idiot box," "the one-eyed babysitter," "the glass teat," "chewing gum for the eyes," to name just a few.

One of the key formal dimensions of broadcast media that allowed for this kind of rhetoric was radio and television's transmission from one to many. The audience can have its minds programmed by radio or TV only as long as the medium allows for no feedback from the viewer. Video games made it possible to see TV as a participatory rather than a one-way medium, and this contrast tapped into a line of thinking about mass society during the era of protest and counterculture of the 1960s. Inspired by C. Wright Mills, the student movement of the New Left coined the term "participatory democracy," an alternative to representative democracy in which all members of a community discuss and deliberate about policy until achieving consensus.[14] Mills's populism decried the dominance of American society by a tyranny of elites, and the students protesting against corporate power, militarism, and civil rights violations saw society's institutions as unprepared to undergo real change. The 1962 Port Huron Statement, a manifesto of the Students for a Democratic Society, urged: "Let the individual share in those social decisions determining the quality and direction of his life." Politics would bring the individual "out of isolation and into community."[15] Participatory democracy was a means to this end. In the historical account of Milton Viorst, "Participatory democracy would overcome a sense of powerlessness by having people partake directly in making the decisions that affected their destiny."[16] This was presented as a solution to the failings of mass society. If mass media imposed conformity of thought through its homogenizing message, participatory democracy would engage individuals on their own terms as political subjects.

In the 1960s and '70s, many changes in TV were underway that were supposed to promise the redemption of television from its status as mass media within a mass society through improvement of its technology and forms. Videotape decks and cameras, cable and satellite television, video art, and possibilities for two-way television permitting audience feedback and participation were all topics of regular discussion in the popular and trade press. So was the futuristic possibility of computerizing television, turning the domestic set into a multimedia terminal networked to news, libraries, shops, offices, and other information sources, as well as other users.[17] These technological developments were presented as revolutionary solutions to some of TV's problems, and as possibilities for television's democratization. Ideas of improvement and democratization of TV at this time were premised on the same shifts that we find in ideas about media convergence beginning in the 1990s: of media communication from one-way to interactive, and from audiences as passive consumers to active users. With cable TV, for instance, viewers would be free to choose programs that would really suit their interests rather than being subject to whatever the networks and their sponsors expected to be profitable. Some of these cable programs might not even be beholden to sponsors for their funding. With video technologies, people would be able to borrow or rent tapes of various kinds or record programs off the air to view at their convenience. The empowerment of audiences is figured in such scenarios as a contrast to network television viewers, who are regarded as victims of mass communication.

Video games allowing users to make the television set into an active, participatory medium taking input from the viewer/player were hardly a politically revolutionary intervention in popular culture. Surely they did little to diminish the power of mass media. But they did appear to many observers—and were presented by the marketing discourses of the games companies—as an improvement on the medium of television, which would solve some of the problems strongly associated with TV as perhaps the most important single instrument of the mass society. Video games were consistently described as a way to transform the TV set into a device that affords a more participatory experience, a contrast against the typical use of TV sets to watch commercially sponsored network broadcasts. As a form of television technology, video games were presented as a solution to existing media, allowing for the public to engage actively rather than be passive victims. In the Cold War context, this meant that games would suit the democratic personality of the autonomous individual rather than lead to Soviet-style totalitarian control over the populace. This contrast with the old idea of TV as passive advanced an identity for video games as a new kind of television, but also as an antidote to television.

Games as TV

As is generally true of new media, games were offered to the public both as a break from familiar experiences and as an improvement on what was already routine. The language used to describe new media often reveals its remediations of existing media technologies, forms, and practices.[18] Radio was called wireless telegraphy; then television was described as radio with pictures. Email is electronic letters, while the "tube" in YouTube is an old TV set. In various ways, video games remediated pinball, board games, sports from ping pong to auto racing, science-fiction film and literature, and electro-mechanical arcade amusements. Perhaps more than anything else, video games remediated television. The iconography, language, and experience of television were presented as guides for consumers in understanding games and their functions and possibilities, just as the content of early games offered imagery that television audiences were used to seeing, like rocket ships and athletic competitions.

We think of video games as computer games, and often assume that they have always integrated video and computer components. While of course it is true that the mainframe and minicomputer games developed on university campuses in the 1960s were made using computer technologies, many of the earliest games released for public amusement such as *Pong* were made entirely using television engineering skill and materials. As Henry Lowood argues, "The reading of *Pong* as a product of the computer age sidesteps the emergence of the videogame out of TV engineering."[19] The history of both the Atari and Magnavox games is one of innovation in raster video rather than of work adapting computer science to entertainment media. Only later on did home and arcade games incorporate microprocessors and software code; initially the most familiar video games to be played by ordinary people were TV games through and through. This explains why video games were very infrequently called computer games or discussed in relation to computers in the early and mid-1970s. (This changed after a few years.)

They were, however, regarded very often as a form of television. When the brand new Magnavox Odyssey was the secret guest on *What's My Line?* it was presented as a television set on which the guest was doing something the celebrity panel could not see. When the object was finally revealed to the panel, the host, after expressing his amazement at the new device, asked the Maganvox representative appearing on the show, "How do you tune it in?" After connecting the Odyssey to the antenna input and choosing channel 3 or 4, he explained, "You tune it in like any regular program." The language of broadcasting ("tune in") suggests that the game image on the screen is received over the air.

In addition to the term *video game* and its synonyms *tele-game* and *TV game* (as well sometimes as *home TV game* to distinguish against coin-operated games in public places), the language used to describe early games was frequently more elaborately televisual. A description of the first home game console in a 1972 *New York Times* article referred to a Magnavox Odyssey game as a "broadcast."[20] A *Popular Mechanics* feature in 1976 described a sports game as follows: "The playing field is just another program you tune instead of comedies and commercials."[21] *Radio-Electronics* called TV games "a new kind of entertainment being offered on [the] home TV screen."[22] *Consumer Reports* explained, "*Odyssey* uses your TV set, including the picture tube, just as TV reception does."[23] *Business Week* in 1975 called electronic games "TV's Hot New Star."[24] The packaging of Atari's console version of *Pong* made prominent the phrase "for your home TV." As Sheila Murphy argues, the small screen "served as a stable and familiar referent for consumers and users who were first learning to read the semiotics of … video game systems being connected to the more recognizable television set."[25] The language used to describe the new games supports this notion.

Ralph Baer, who invented the Brown Box that was the model for the Odyssey as well as an inspiration for *Pong*, initially called his idea for this invention "Channel LP—let's play!"[26] A video game console produced by Fairchild in 1976 was known as Channel F. A sales brochure for Channel F used phrases that link games with the TV set on which they would be played, such as "Now Playing" and "Stay Tuned." Mattel's Intellivision console, short for "intelligent television," organized its games into thematic "networks" such as sports, action, strategy, arcade, and children's learning. Above an Odyssey game pictured in a Sears catalog was the excited description: "All the action takes place on your TV!" This might have caused some confusion, however. Sears catalogs in the mid-1970s regularly cautioned prospective buyers of video game consoles: "TV not included." One wonders what expectations consumers might have had if they needed to be told this. Perhaps some people might have thought the pictured product *was* a television set that plays games in addition to receiving broadcasts.

Figure 2.1
Fairchild Channel F brochure.

Imagery of television was central to the representation of games in visual media such as magazines and catalogs. The most typical image was of players, often opposite-sex couples or parent–child combinations, posed in front of a TV set in a comfortable living or family room decorated with wall-to-wall carpet, couches, and coffee tables. These images would echo the standard image of the family circle in advertisements for televisions and fit games into long-standing discourses of domestic media. The frequent presence of game imagery on higher-end console TVs—larger sets built into wooden cabinetry of traditional styles—establishes clear class connotations for the new technology. It inserts them into the ideal of suburban white

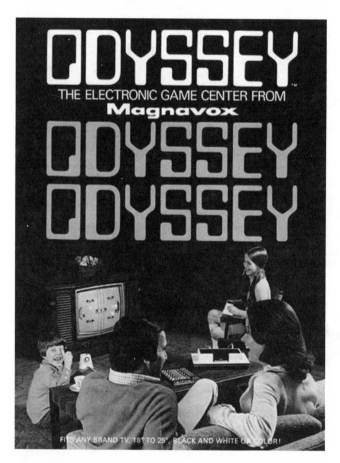

Figure 2.2
Magnavox Odyssey flyer.

American family life in much the same way as traditional representations of television since the late 1940s.[27]

Graphic design representing TV games also borrowed familiar iconography, especially of the bulging rectangle shape of television. The "electronic tele-games" portion of a Sears catalog set the "tele" portion of the text against a bulging rectangle with rounded corners, emphasizing the hardware used to play the new device. Game catalogs for Atari, Intellivision, and other programmable consoles (i.e., consoles that would take cartridges to play different games) typically represented the screen image within the same shape to make clear how they would look on a TV set.

The identity of video games with a TV set is especially clearly expressed by a toy produced by the Louis Marx & Co. for sale in 1974 called "T.V. Tennis." Clearly inspired by *Pong*, Odyssey *Tennis*, and numerous similar games coming to market at the time, this electro-mechanical plastic toy with no video components takes the form of a green CRT set with control dials in the corners, mimicking the design and functionality of *Pong* without its electronic hardware. A video game sans video, T.V. Tennis reveals how the cachet of the image of television, and the novelty of interacting with imagery shown on a CRT display, helped establish an identity for the new technology of the video game in the early 1970s.

Figure 2.3
Catalog detail: "tele-games" from the Sears Wish Book for the 1979 Holiday Season.

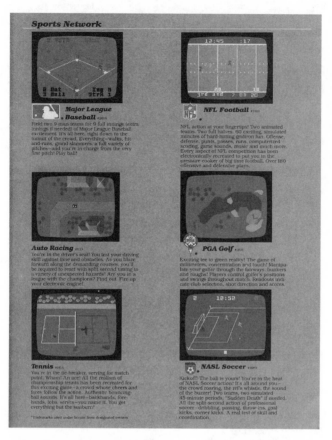

Figure 2.4
Intellivision catalog.

New Tricks Your TV Can Do

Early video games might have relied on TV to establish an identity, but TV's identity was hardly stable at this time. At least as important as games' dependence on a traditional conception of TV was their prospect for transforming TV, making television do new and different things. This occurred during a time of media change as television's status was in flux. "The widespread public acceptance and use of home video game systems by a broader audience," argues Murphy, "indicates that consumers were rethinking television's role as a home technology in the mid-1970s."[28] Video games were among a cluster of emerging video technologies that promised to renew television by opening it up to new forms of programming and new modes

Figure 2.5
Marx T.V. Tennis game.

of engagement. In addition to games, the two most notable developments in television technology of this era were videotape becoming available outside of industry, and cable television subscription services launching as an alternative and supplement to broadcasting. Video games were understood as a new television technology, like video and cable, which would change the terms by which TV was experienced, and thereby also change TV's social and cultural meanings.

In assessments of their impact on the television industry in the middle of the 1970s, video games were often discussed in tandem with consumer videotape recorder decks, which at the time were still called VTRs. These assessments might appear in industry trade papers such as *Broadcasting*, which in 1977 described video games and VTRs as "the two big TV toys," but they were also circulating in publications aimed more at general readers.[29] Video games and VTRs had a number of things in common. They were both peripheral devices that would plug into the television's antenna or CATV input and substitute new content for the broadcast or cable TV signal. They both represented competition for the attention of television audiences who might turn to them as an alternative to the offerings on their local stations or, in some cases, cable channels. And they both implied a new conception of the television user as a more active agent, choosing content at will and no longer beholden to network programmers and their schedule grid.

Millions more video games were sold in the mid-1970s than VTRs, though this was a matter more of supply and price than of consumer interest.[30] Of the two, tape had been seen as the more potentially revolutionizing force in television, promising to open up video content well beyond safe and formulaic network fare. "The Video Revolution," a 1970 *Saturday Review* story looking forward to home video, promised that through the new technology, the TV viewer would be "rescued" and "liberated" after decades of "being held captive by network and sponsor." Videotapes would offer "salvation" in the form of "a dramatically wider range of choice, and an adaptation of program to your convenience."[31] In the hopes of its champions, video would bring high culture, instructional or educational programs, Hollywood films, and children's shows to audiences who were presumed fed up with the commercial TV status quo.[32] In the 1960s and '70s, American cultural critics often speculated on the promise of home video to be a form of cultural uplift, giving the mass audience better forms of programming in diverse forms and genres. They were eager to see the oligopoly of the broadcast networks blasted apart by new channels of distribution for video content. They presented this as a democratization of mass media, though elite taste would still govern the available choices.[33] Highbrow critics were skeptical that anyone would want to use home video recorders to save most television programs for later viewing, and saw the new technology offering a departure from the standard formats and experiences of television. But when Sony released its Betamax deck in 1975, its advertising pitch to consumers was "Watch Whatever Whenever," giving the TV audience agency to program their own schedule. The visual rhetoric of the Sony campaign was similar to that of video game console advertising, with the device pictured before a TV set, linking them as components of a new technological ensemble, and suggesting a revised conception of television. Whether offering alternative fare or viewer agency, the VTR was perceived as a source of disruption to TV's institutions and formats.

From the TV industry's perspective, video games and videotape were a pair of related threats to their business. An *Advertising Age* story on the two in 1977 makes this clear in its headline: "Innovations in Video—Nightmare for Networks?" It begins: "Videotape recorders and video games may be the

Figure 2.6
Sony Betamax advertisement: "Watch Whatever Whenever."

two biggest monkey wrenches ever thrown at the television networks."[34] Among other details, it notes that most video game play occurs during evening "prime time" (when TV networks charge their sponsors the most to air commercials) and that videotape usage shifts viewing away from evenings as well. Both of these new devices upset the TV networks' expectations of patterns of use of the television set. VTRs would allow for skipping commercials, while video games are not an advertising-supported format, joining them together as an obstacle between the television networks and consumers whose attention they desired. In a 1972 story, *Newsweek* quoted

Bobby Sherman, a writer, claiming he plays Odyssey games during commercial breaks while watching television, sometimes continuing through a few minutes of a show if "the game gets hot."[35] The idea that television sets could be used for popular entertainment that does not include advertisements was a novelty, and the *Advertising Age* reader would naturally regard this with some apprehension.

Changing Times, a general interest magazine, put this same development in more positive terms in a 1976 article. While Madison Avenue and the TV networks construed new video technologies as a potential nightmare, ordinary consumers were invited to greet them as exciting, liberating developments. Explaining the headline "New Tricks Your TV Can Do," the article lists electronic games and videotape, as well as projection television sets with their "theater big-screen effect." The lead paragraphs portray a bold turning point in television history at that moment: "Your TV set has been fundamentally the same from the start—a box that allows you to see what someone else wants you to see whenever that someone decides to let you see it. Now television is up to new electronic tricks long dreamed of, long discussed and finally here."[36] Watching on videotape is described as "doing your own TV programming," and all of these new developments in TV technology promise to break viewers out of the broadcast mode of television set usage, whether by playing a game, watching on their own time, or making the home into a cinema-like space. By aligning video games with videotape and big-screen TV, *Changing Times* packaged them all together as new media upsetting the stability of television's cultural status.

Cable television was less often linked with video games in the 1970s, though there were several plans to offer games through cable subscriptions.[37] Cable, however, was another key example of a new television technology promising to break the network oligopoly and open up TV to new, diverse, and worthwhile experiences. As with videotape, expressions of hope for cable TV in the 1960s tended toward bright optimism, regularly offering a utopian sense of revolution through technology. One commentator in 1972 proclaimed, "An almost religious faith in cable television has sprung up in the United States," ascribing this to cable's potential for offering "a way out of the vast wasteland of commercial network television."[38] Like videotape, cable subscription would multiply the options available to viewers previously given the limited choice of three major networks and a handful of local stations. The scarcity of TV programs was viewed as an artificial condition in need of remedy by cable television's multiplicity of options.[39] Cable also had the potential to break the "one-way" nature of broadcasting and offer new "two-way" communication that would empower audiences to speak back to the TV set. The discourse of cable television in the years before it was widely offered was one of transformation from passive mass communication to audience participation. TV itself was constructed as a problem in need of a solution, and cable subscriptions represented a solution from outside the TV industry and the system of commercial network broadcasting. Thomas Streeter writes that cable television was seen as a technology to "empower the ... passive audience and eliminate the one-way quality of TV."[40]

Video games emerged against the background of these discourses, in which the trouble with television is always seen as being inherent in the one-way nature of commercial broadcasting as a form of mass communication, and in its limited range of options. The way forward to a better future has always been to open up TV to the participation of its previously passive viewer in diverse new experiences. This was the promise not only of cable TV and videotape, but also of electronic games like the Odyssey and *Pong*.

Video against Television

Video games, along with VTRs and cable TV, were part of a renewal of the status of television in the 1970s. But these new media technologies were also challenges to television, undermining its functioning industrially and culturally. As ways of using the TV set apart from the institutions of television broadcasting (e.g., networks and stations), the new video technologies of the 1970s were opposed to TV in its traditional conception. Video games were in one sense a "trick your TV can do"—a new way of using TV that would improve its value. In a more critical sense, however, video games were a trick at TV's expense. At the same time that games were making TV better, they were also providing an alternative to television that would reveal TV's flaws in clear focus.

One way to think of the changes television was going through in the 1970s is as a "makeover."[41] A makeover is premised on a "before" identity being unacceptable or even shameful. Television's perceived failures have historically necessitated technological fixes offered by new audio-visual media. Sheila Murphy sees the emergence of video games in particular as an early instance of television's encounter with digital media—what in the early 1980s was called the marriage of computers and television, and what later would more simply be described as convergence.[42] TV in its old identity as a broadcast receiver has often been seen as the component of convergence in need of improvement or amelioration; by this logic, television is a problem that convergence solves.[43] We see this logic at play in discourses of the 1970s figuring television as a problem and the video game console as the solution.

Some of the most negative accounts of television in video game discourses were articulated by the "fathers" of the new medium, Ralph Baer and Nolan Bushnell, as explanations for their motivations as inventors of electronic game devices. Baer, an engineer, had been interested in devising a game to build into television sets in the early 1950s, but his employer was unwilling to pursue this research and development. When Baer conceived the plans that became the Odyssey in the mid-1960s, he has said, he was motivated by the desire to use television sets for a new purpose. Baer wanted to offer something to do with a TV "other than watch stupid network programs."[44] He has said that the many millions of television sets in use in the 1960s were "practically begging to be used for something other than watching commercial television broadcasts!"[45] Bushnell, the Atari founder sometimes described in the 1970s as "King Pong" and frequently profiled in the press, told the New York Times in 1978 that video games represent "the first time people have been able to talk back to their television set, and make it do what they want it to do. It gives you a sense of control, whereas before all you could do was sit and switch channels."[46] Bushnell told a magazine interviewer in 1974 that he wanted to make amusements be "not just spectator-oriented but participatory."[47] Rather than expressions of their individual desires, we can see these utterances as social ideals of media and technology speaking through Baer and Bushnell. These men were merely expressing their culture's commonplace ideas of television's shortcomings, but, especially in Bushnell's case as a more public persona, giving these notions the symbolic weight that comes with being a prominent innovator in both technology and culture.

The contrast of passivity and participation was standard rhetoric in the 1970s, frequently appearing in popular and trade press accounts of new media. An article in Sales & Marketing Management in 1976 promised, "With the video games, it's strictly a participation sport."[48] Video games were presented as a more active, engaging, and exciting use for the television set than merely watching it. But this idea is premised on the failure of TV in its usual conception to be anything more than a time-waster. The implicit indictment of television as a key component of the mass society, as a force promoting idiocy, conformity, and civic disengagement, is the background to any such appreciation of new media as a positive alternative to broadcasting.

Popular press stories introducing new technologies upon their public release, assuming a positive slant, are typically breathless and not infrequently techno-utopian. The stories hyping TV games in the 1970s fit this bill. Their optimistic descriptions of the new devices reveal as much about attitudes toward TV as about video games. In some instances, the games are presented as a revolutionary change for TV. In a *New York Times* story about the release of the Odyssey in May 1972, for example, the president of Magnavox (a long-time manufacturer of television sets) is quoted claiming his company's product "is an educational and entertainment tool that transfers television from passive to an active medium."[49] TV commercials seen at the time of the product's release promised it would give "a new dimension for your television." A *New York Times Magazine* feature two years later, under the headline "The Space-Age Pinball Machine," included quite similar language:

> Odyssey costs $99.95 and is simple enough for the average consumer to attach to the back of his TV set, transforming the otherwise passive box in his living room into a game screen with two "player" blips and a "ball" blip moving across it.[50]

The language of *passivity* and *activity* is frequently matched in these descriptions with terms similarly describing agency, such as *engage* and *control*, as a means of distinction between games and ordinary TV. The article first describing TV games for readers of *Time* magazine in 1972 paints the following contrast:

> The average American spends six hours a day gazing passively at television. Soon he will have an opportunity to play a more active role in what appears on the screen of his set. Last week the Magnavox Co. demonstrated a device that will give set owners a chance to engage in electronic table tennis, hockey, target shooting and other competitive games on their TV screens.[51]

The hobbyist-oriented magazine *Radio-Electronics* placed games into a similar before/after scenario, representing a transformation of television from meager entertainment to a more elaborate fantasy world of immersive sporting simulation:

> For a considerable number of years, we sat in front of our TV sets and let them entertain us with moving pictures on that little screen. ... Yet there is a new kind of entertainment on that home TV screen—it's a Ping-Pong game, a soccer field, a shooting gallery and others and you, who until now have been a passive viewer get to control the action.[52]

In all of these accounts of the new technology, we find that the television is not merely used in a new way, but the user's experience is understood to have a value diametrically opposed to the typical television audience's. Another hobbyist publication, *Mechanix Illustrated*, captures this totally different idea of the television audience, a fresh identity in contrast to the lazy spectator of the past:

> TV screens used to be just for watching whatever the network or the local station felt like putting on the air. Now the home set has become the center of family sportsmanship. ... It appears that the bouncing blip could change the habits of American TV gluttons. It will surely get them more involved.[53]

The unconcealed disdain for television in this writing, and for the conditions of television viewing in the typical American family's home environment, speaks particularly to the gender and class politics of new media discourses of passivity and activity. In the move from "just ... watching whatever" as "TV gluttons" to being a "center of ... sportsmanship" for those "more involved," this writer taps into a long-standing characterization of TV as feminized mass culture, and of interactive new media as a more masculine and legitimate form of leisure-time recreation.

No concept expresses these ideas about old and new media in the 1970s as potently as *participation*, an unambiguous virtue in any writing of the period concerning media. Reporting on the popularity of electronic games on Christmas of 1975, *The New York Times* quoted a retail analyst asserting: "We may be leaving the spectator era for the participation era."[54] This new era was signaled not only by the games played on a TV set, but also by the CB radio craze and the growing popularity of physical exercise such as jogging. The analyst quoted in the *Times* claimed that thanks to participation replacing spectatorship, that year he was even planning to skip watching the Super Bowl.

Video games would thus be part of a new turn toward more active and social leisure experience. A first-person account of an encounter with video games in the highbrow *Saturday Review* in 1977 by the magazine's editor and publisher, Carll Tucker, lamented the effects of television on personal interaction, and the time wasted by the national habit of watching TV every evening. The author sounds a familiar theme, evoking notions of mass society and participatory democracy, in relating his first encounter with a video game:

> Instead of merely talking at me, the television screen was challenging me, inviting me to participate. Moreover, the television was proving a social medium, a playground where two people could meet and relate to each other. For a change, it mattered whether or not I was paying attention to what was taking place on screen. ...

Television isolated us from each other; it collected the world into a global village and locked each of the villagers in cells. *Pong* obviously is not the solution to this, but it is a hopeful sign. If, for a few hours, instead of the isolating TV fare, you have a playing field where live (as opposed to "live") people deal with each other, communicate, the world is that much more intimate and real.[55]

With participation comes a renewal of the television audience's sociability and interpersonal communication in place of mass communication. Early video games like *Pong* and most Odyssey games were two-player affairs, like the racquet sports from which they were adapted. Whether one or two played, the game console was always pictured and described at this time as a group rather than a solitary activity. Representations in magazine items, advertising, and catalogs showed two or more players at the television set. This participation was thus not only between individuals and media, but also among players.

A similar point is made in *Pilgrim in the Microworld*, a 1983 volume published by Warner Books (owned by Atari's parent company). The author, the sociologist and pianist David Sudnow, began to play the Atari game *Breakout* and became so fascinated by his experience that he wrote a study of games addressing the general reader. Sudnow's earlier book had been a finely detailed analysis of the experience of playing piano, and his intricate descriptions of his *Breakout* sessions apply a similar expository style to electronic play. This legitimates video games by treating them as a cognitively engaging, even expressive medium in which the body and mind have to learn to work together. Sudnow is no television viewer; he sees Atari as a quite different type of technology from the broadcasting receiver into which the console is plugged. Sudnow has to buy a TV set to play Atari—he is intellectually above television—and his assessment of Atari's value is in relation to television's: "Bless you, Atari ... you've resocialized us after thirty years of being vaguely with each other during prime time."[56]

This sense of new video technology being characterized, in distinction to broadcast television, by participation was not a new idea; its origins were in already existing discourses about television and new media of the 1960s and early '70s. Jason Wilson draws connections between Baer's and Bushnell's ideas about video games as a more participatory form of media with the objectives of new media artists of the 1960s using television as a medium of experimental artwork. Video art of the 1960s by Nam June Paik in particular and video games of the 1970s were alike in their aspirations to transform television as it is experienced, making it more productive and changing the viewer's role into one of engagement and purpose rather than mere spectatorship. Paik's television work opposed the regime of network broadcasting as one-way mass communication, opening up the technology of video to manipulation as electronic signal and output, making the TV set into a plastic medium for the artist's and audience's creative manipulation. For instance, his 1963 work *Participation TV* invites the spectator to use a microphone and sound frequency amplifier as an input for the electronic signal producing abstract imagery on the CRT screen. Paik's statements of intention from the 1960s are remarkably similar to popular press accounts of video games in the 1970s, contrasting passivity with activity and one-way communication of mass media with a more reciprocal relation between the audience and the medium.[57] Both video art and video games are thus premised on participation in distinction to broadcasting. Both invite "a new kind of productive spectatorship."[58] The 1969 video art exhibition in New York City called "TV as a Creative Medium," which attracted the notice of the popular and alternative press, collected a group of avant-garde artists who, Marita Sturkin explains, regarded video in distinction to commercial broadcasting, "as viewer participation, a spiritual and meditative experience, a mirror, an electronic palette, a kinetic sculpture, or a cultural machine to be deconstructed."[59] According to Wilson, *Pong* was based on an idea similar to one energizing a contemporaneous new media avant-garde: making television "manipulable" by its users.[60]

This idea of a manipulable television, of the audience as active agents making graphics move around the television set, implies a transformation not only in the address of media to audiences, but also in the orientation of audiences to media. The language of transformation suffuses discussions of early video games in the home. TV sets, according to some accounts, were being turned into "electronic game boards," a nice image of remediation combining the interactive play of board games with the advanced technology of electronic media.[61]

The packaging, marketing, and advertising of video games consoles are perhaps the most powerful instances of this rhetoric of transformation. Intellivision was sold by unpacking its portmanteau name combining the words "intelligent" and "television." These words were printed on the box in which the console was sold. In commercials starring the erudite George Plimpton, speaking in his classy mid-Atlantic accent, Intellivision presented itself as the sophisticated game system for discerning players, showing side-by-side comparisons with Atari to emphasize Intellivision's graphical superiority. The tag line was "This is intelligent television!" While "intelligent television" might have been an oxymoron to cultural critics of the pre-video era hopefully anticipating the medium's liberation by new technology, by the late 1970s and early '80s the idea of television's shift from stupid to smart through audience participation had become familiar.

Atari's print and TV campaign in the later 1970s also sold video games as transformative, marking television's passage from the passivity of mass communication to the participation of video. Video games would invite players to act out fantasies of masculine empowerment. TV commercials featured famous male athletes whose skills on the field would comically fail to translate into video games. A baseball hero stands with bat raised at the plate and looking in the camera's direction says, "Okay Atari, let's see your best pitch!" Cut to the umpire: "You're out, Rose!" This humorous scenario reinforces the activity of video game play by association with athletics, and also empowers a perhaps youthful player to defeat the pros in video simulations of real-life contests. The Brazilian superstar of the New York Cosmos confesses, "I quit soccer to play Atari," but in the next shot a little girl, perhaps his daughter, waves a finger at him delivering her line in a sing-song voice, "You need more practice, Pelé!" The consumer is invited to develop a skill by playing games, just as one would with a serious sport like soccer or baseball. At the end of the thirty-second spot, a voice-over reminds consumers that the action in Atari would be "on your own TV set." Its tag line is "Don't watch television tonight. Play it!" The same phrase appeared in Atari's print ads, defining Atari's identity in relation to TV and promising the transfer of television from a passive to an active medium. In messages like these, the public image of video games was of something you do on your own TV set, but also a means of making your TV set do new and exciting things. To "play TV" is to redress the failure of mass media and substitute a new sociality, with its promising dimensions of purposive engagement and user control, for the old experience of television as a feminized, mass-culture, broadcast medium.

At Play in the Home Entertainment Center

As a consequence of its renewal through technological change, TV's place in popular imagination shifted during the years of early games. No longer the same medium it had been in the 1960s, television had to accommodate its newly participatory and diverse potentials. Home entertainment increasingly included the television set as one of a cluster of technologies including videotape machines, game consoles, cable boxes, and later computers that used a CRT display as a monitor, along with the TV set speakers for sound. Stereos, including radios, audio cassette decks, and record players, and slide projectors might also be included in such an ensemble, making

Figure 2.7
Atari advertisement: "Don't Watch TV Tonight. Play It!"

it all the more multimedia. The panoply of options for electronic home entertainment would stand in contrast to the scarcity of channels available when the home screen was only a broadcasting receiver. As *Electronic News* reported in 1977, "The television set is evolving into the hub of a complex of entertainment and information functions." One primary indicator of this was the rise of participatory play: "The emergence of the video game … is being viewed as a harbinger of the home entertainment center."[62] By this time, millions of video games had been sold for home use, and their presence in so many homes was changing perceptions, particularly in electronics industries but also among the general public, of television's purposes and possibilities.

The Magnavox company is a good example of a purveyor of a convergent experience of home entertainment. As a television set manufacturer that produced the first home console, Magnavox was central to the development of the expanded conception of TV as a media hub. Its 1978 advertising aimed at retailers pictured its most important products clustered together on a delivery truck. Included among the items for sale in Magnavox showrooms were traditional console television sets and wooden cabinetry. In addition, however, Magnavox offered videotape players and cameras, hi-fi components, and portable radios. Pictured in the center of the image was its new game console, the Odyssey^2, a "programmable" machine to rival Atari's VCS released in 1977. Here we have media convergence before convergence, a home entertainment center clustered into one shipment of hardware. Video games were at the center of this new image of home entertainment, displacing TV from the focal point of Magnavox's sales effort.

Electronics manufacturers and retailers might have greeted video games, among other technologies, as a way of making TV into a "home entertainment center," but within this center the components still needed to be understood on their own terms. Representations of video games in the late 1970s and early '80s often expressed anxieties about the status of television and games in this newly renegotiated media context. In particular, television's residual status as passive and games' emergent status as active appear to clash when it seems that the use of one medium by the other complicates these media ideals. For instance, a cartoon in *Changing Times* from 1978 pictures an opposite-sex couple sitting on a sofa dressed for tennis, with racquets on the floor and leaning against the furniture. But instead of playing tennis outdoors on a court, they are playing a *Pong*-like video game version. This image seems to question the new common sense of video games being active compared with television, poking fun at those claiming such transformational participatory experience for TV game players. In this representation, what the players are doing looks more like watching TV than playing sports; thus the silliness of their attire and equipment.

A similar anxiety is expressed in a four-frame comic in *Blip*, a short-lived periodical published by Marvel during the early 1980s video game craze. Big brother is playing an Atari-type game while little sister observes. She asks, "When do the commercials come on," and he answers, "There aren't any." Already we see a humorous confusion between games and television, and a potentially more virtuous quality of games by comparison with TV: their lack of ads. But she misunderstands: "You mean like *Sesame Street*?" He

Figure 2.8
Changing Times, 1978, showing the tension between games as TV and as participatory activity.

answers, "Yeah, something like that." In the final punch-line panel the girl goes to her mother in the kitchen, where mom asks, "What's your brother doing?" Her reply: "Watching educational television." The joke is partly at video games' expense, as a medium already gaining a reputation for being violent and destructive, though some claimed benefits of playing for some kinds of learning (an idea to return in chapter 5). The identity of the gamer as the son rather than the daughter reinforces the cultural stereotype of the video game player as youthful and masculine, and positions play at once as a more active use of TV than watching television shows, and as a less socially productive use than watching the best of television programming for young people. But perhaps most interestingly this is an expression of the unstable cultural identity of TV: with video games, it's no longer clear what television is or will be.

These images show the extent to which, as a remediation of television, the video game of the 1970s and early '80s was at once a break from the past and a continuation of it. It offered amusement in front of the home screen, in familiar family spaces using the same TV sets that had been regular leisure-time companions for two or three decades. It also promised a departure from television and an improvement on its experience. For those who encountered games in these early years, the meanings of the new medium

Figure 2.9
A *Blip* comic strip, positioning games between conceptions of good and bad television uses.

would very likely have been tied up with ideas about the television set and about television's institutions, forms, and practices. To the extent that video games were an alternative to TV, they found an identity in relation to a medium long denigrated on the basis of its feminized and lower-class cultural status. This reputation would be informed by notions of television's power over an audience powerless to resist its manipulation and control. The identity of video games as a masculine, youthful, active, and engaging medium of play originates, in part, in this distinction: in place of the old, disreputable mass medium, television could be a platform for participation.

3 Space Invaders: Masculine Play in the Media Room

Did you play with a friend on a rainy day?
Did you play with your dad?
Did you show him the way?
Did you play with your sis?
Did your mom always miss?
Did you play a game from Atari?
Have you played Atari today?
—TV commercial, late 1970s

The TV game that emerged in the early 1970s with *Pong*, Odyssey, and many other ball-and-paddle consoles, was by the later years of the decade becoming a fixture in the homes of American children and teenagers, such as those depicted in this cheerful Atari commercial. It presents members of a "typical" (white, middle-class, suburban) family engaged in sociable play around the console. The scenario is one of family togetherness, but the targeted consumer is clearly one member of this group: the son. The boy addressed by the ad is an ambassador of gaming, teaching his father to play, making room for his sister on the couch when a friend isn't around, laughing with his mother over her failure to master the device. The ideas about video games offered at this time in television commercials, as well as department store catalogs, newspapers and magazines, movies, and a burgeoning fan culture, helped establish an identity not just for the medium but also for its typical users. "Anyone can get hooked," the ad concludes, meaning that even a woman—a mother—could be susceptible to the product's appeal. But the message is clear. Even though they are played in the family room, where they bring together participants of different ages and genders, video games are especially a boy's amusement.

In the later 1970s, video game commerce and culture expanded at the same time that games became more centrally identified with youth and

Figure 3.1
Atari commercial, "Have you played a game from Atari?"

masculinity. The more technically advanced game consoles released in the second half of the 1970s, known as the "second generation" of game hardware, included Fairchild's Channel F (1976), the Atari VCS of this TV spot (1977), and Mattel's Intellivision (1979). These were "programmable" consoles with game cartridges, whose quality and variety would expand the time and interest for play. Now a game could be contained in a single chip within a cartridge sold separately, rather than hardwired into the game console. TV games were sold and consumed not just as hardware but also as software, becoming commodities separate from the consoles that accepted them, and marketed as specific games rather than as devices to transform the television set into something new. The games themselves, in addition to the console device, were the objects for sale. And the sales pitch in ads for specific games like *Berserk* or Atari's *Basketball* would include the same kind of representation as "Did you play a game from Atari?"—mixed genders or ages playing in the common space of a well-off family home, but also an address to young, male consumers.

Like *Pong*, many of the popular games for home play in the era of the first programmables were ports or imitations of well-known arcade titles. Many also had an aggressive, militaristic quality. *Space Invaders* was the biggest hit of these years for both coin-operated and home games. An Atari console could play *Space Invaders*, as well as *Asteroids*, *Missile Command*, *Defender*, and dozens of other video games, many of them familiar from arcades. Some of these products—consoles and games—became significant commercial successes and claimed more and more of Americans' leisure time. As of 1976, Atari was part of Warner Communications Inc., and within a few years was earning the parent company a substantial amount of its revenue.[1] Thus, in the first decade of their availability for play by the general public, video games went from being an unfamiliar innovation in television and electronics to a powerful economic and cultural force, part of most young people's lives and of many older people's. The arcade, home console, and handheld electronic games were also joined in the later years of this period by home computer versions, as electronic play proliferated and grew more varied and sophisticated.

Video games' identity was established within a cluster of contexts. The arcade and home of the 1970s and early '80s were distinct spaces, and in some ways opposites. In popular imagination, as we have seen, the arcade was a potentially threatening destination frequented especially by teenage boys. It was associated with pinball, an amusement with a historically low cultural reputation associated with gamblers and crime, situated within a masculinized public sphere unless banned, as it was in many cities.[2] The home, by contrast, was idealized as a sanctuary of safety and comfort for the family, a feminized private sphere. But the two places could blur into one another as arcade games were reproduced for the home screen, and as home gameplay was offered as an experience similar to—or superior to— play in an arcade. Commercials for video games in these years often presented console games offering an "arcade experience." Coleco ran an ad in the early 1980s in which a young male player is transported from the arcade to the home while playing a ColecoVision game in which spaceships emerge from the screen and hurtle forward, relocating play midgame from a public to a private venue. The player is so immersed that he does not notice that he has been moved into a living room while he plays. The voice-over promises "arcade controls and arcade graphics that let you have the arcade experience at home." Rather than opposed, we might think of arcade and home games as a pairing like the movie theater and videotape deck, both offering experiences of the same kind but using distinct technologies, encouraging distinct practices, and defining distinct spaces. Ultimately the home consoles were to claim the majority of young people's time and attention for video games, and were most central to the rise and fall of companies in the electronic games industry in this early stage.

As movies were with television and videotape, video games in the later 1970s and early '80s were *domesticated*, a term with two related senses.[3] Domestication of media is, first of all, akin to the taming of wild animals. Video games had to be made safe, familiar, and predictable. This kind of domestication is the process of new technologies becoming integrated into everyday life and passing from novelty to regular usage. Domestication also refers to the literal integration of games into domestic space, the space of the home, and in particular the idealized single-family home of white suburban America during the Cold War. The taming and familiarizing of games and the incorporation of games into routines of middle-class family life were part of the same process of video games coming into an identity as a medium with widely shared and stable meanings and purposes.

Thinking of the early history of video games in terms of domestication prompts the following contradiction. In the promotional discourses of the game companies, the console was to be a fixture of the recreation room, an instrument for bringing a middle-class family together in active, social play. Like television and radio before them, video games were often represented as the focal point of a family circle, a new electronic hearth. Like other forms of home recreation, such as playing cards, board games, and table tennis, video games were a way for the family to pass leisure time together, and also to entertain friends. However, video games drew extensively in their forms and representations on traditions of masculine play and boy culture, and offered a form of recreation and leisure quite at odds with the ideal of the family circle in feminized domestic space. A great many of the video games from the medium's first decade include some element or combination of sport, space adventure, and combat.

This masculine character of video games would always be in tension with the settings in which games were typically experienced. Rather than integrating them harmoniously into family space, we might see video games in the home as an escape for children out of the domestic realm. This escape would offer a world of play in virtual spaces, and in particular a form of play deriving from a history of boy culture that resists the middle-class propriety and companionate leisure of suburban American ideology. The contradiction, then, is one of age and especially gender. Video games were at once an activity for families to do together and a way for the family's children, and particularly its sons and their male friends, to experience their own modes of adventurous, competitive, and often violent play within domestic spaces from which otherwise they might have more literally escaped.

The space of the home gave meaning to video games, in some ways similar and in others different from the meanings produced through the space of the public arcade. The consumption of media is always shaped by the contours of their location, by the social organization of space.[4] Representations of games and consoles in popular discourses of the time, as well as the forms of the consoles, cartridges, and games themselves, produced and circulated ideas not just about a new medium of video games, but also about masculinity and femininity, youth and adulthood, family and home. Representations of home video games, including magazine articles, retail catalogs, print and television advertisements, game cartridge and box art, and the forms and content of games themselves, offer traces of the domestication of the medium. So do secondary sources such as social scientific studies of early video games in the home, sources that reveal conceptions of home leisure and the gendering of play in this period and historically. The domestic material culture and interior geography of early games have much to teach us about the medium's emerging identity.

Play in Modern Domestic Space

Video game consoles like Odyssey, Atari, and Intellivision were played in all kinds of locations, including urban and suburban houses and apartments of many socioeconomic levels. They were idealized, however, similarly to many other consumer commodities of the twentieth century, by being represented in a great many instances in the bourgeois domestic landscape of the suburban single-family home. Like television in the 1950s, video games in the 1970s were sold to the public on the basis of advertising and promotional imagery of well-off white families coming together around the new medium as earlier generations might have gathered around a piano or fireplace. A pair of promotional images for Magnavox products (see figs. 2.2, 3.2) makes clear the semiotic linkage between TV and TV games in popular imagination: in an advertisement for a TV set from the 1950s and another for the Odyssey from the 1970s, we see remarkably consistent iconography. The details but not the underlying meanings have adapted to the style of the times, but both images are of an affluent, white, nuclear family circle brought together by electronics. Both images even represent the family enjoying a sporting event, though in the new version the competition is in the living room. In popular press imagery and catalog photography, as well as in game advertisements, the uses of the console in the home are frequently those of middle-class white family members of mixed ages and genders. One type of domestic space was especially important for the new medium of the video game, a room that would be known by a number of related names including family room, recreation room, rumpus room, or media room. In identifying the meanings and practices most directly associated with video games, it will be valuable to know something about this room and the practices and meanings identified with it.

Ideas about the private spaces of the home shape the meanings of the technologies used in those spaces. New media for domestic use, such as video games in the 1970s and early '80s, emerge into a context of mediated leisure and work within or against the range of meanings already in place for leisure and media. As Erkki Huhtamo observes, "Domestic media, including video games, are intimately connected with the rituals and practices that constitute domesticity."[5] Video games were not only media, however, but also forms of play. Their significance in domestic space draws upon already established practices of watching television and playing board games, bridge, pool, or ping pong. It helps to know where in the home video games might have most often been used (or where their use was seen to be appropriate and typical), as the different rooms of a house have distinct meanings associated with identities, practices, and ideals. The design and use of the family home has a history, shifting according to economic, social, cultural, and technological needs. The family home of the 1970s and '80s already functioned under a kind of economy of leisure-time activity,

Figure 3.2
A 1950 television ad by Magnavox.

with certain identities and uses deemed appropriate and desirable for certain spaces.

The history of domestic architecture in the West since the industrial revolution is one of expansion and segmentation of space. It is also a history of changing ideas about domesticity, and the development of the notion of the home as a private, safe, and comforting sanctuary for the family in distinction to a public sphere of masculinized work in a field, factory, or office. A preindustrial home would often be a single room, but advances in heating and building and a separation of public and private conceptions of space and experience produced new configurations of domestic architecture.[6] In Victorian times a middle-class house would include two rooms for leisure time, a front and back parlor. The front room was more outward-facing and ostentatious, showing off the family's finest things and furniture to visitors as an advertisement of the family's status and taste. The back parlor or sitting room was a space associated more with comfort and informality, furnished in a more utilitarian and less careful manner. Children would more likely be allowed to occupy the sitting room, and families might retire there after dinner to read or play or make music.[7] The postwar suburban house reproduced this division of space with its more formal living room and more informal family or recreation room, possibly adjacent to the kitchen, or down a flight of stairs in a split-level design. The family room was idealized as a space of companionate leisure to include all members of the household. It was often the place for media technologies such as hi-fi stereo equipment and television, as well as tables for eating or playing cards or board games, and upholstered chairs and sofas suitable for reading and entertaining company. This room would be furnished to accommodate casual entertaining, but its most central and important use would be family togetherness. As one 1970s decorating guide put it, "The best entertaining in the home is the family entertaining itself—which is what the family room is essentially all about."[8]

As the quintessential private space, the idealized house of the postwar American suburb was a key element in the abiding Victorian ideology of the separation of the spheres: masculine and feminine, work and family. As Lynn Spigel argues in *Make Room for TV*, television in the 1950s negotiated between the public and private conceptions of postwar family life.[9] Representations of TV expressed the tensions and contradictions of a society marked at once by gendered spatial distinctions—women at home, men at work—and by efforts to integrate families together within a harmonious domestic sphere, which television was supposed to help accomplish. As a typical place for television viewing by the whole family together, a family room would be central to the spatial configuration of postwar domestic gender ideology, which involved the problematic integration of the male family members into a feminized sphere. As a technology requiring a TV set, video games would naturally be understood in relation to the meanings of television within the home. But also as a kind of competitive, skilled play, as a *game*, the new medium would be understood in the context of another tradition, of family leisure-time amusement in the form of games played by both children and adults.

One trend of twentieth-century home design in particular was to include dedicated spaces for children's play within the single-family home, as well as in private yards or gardens. Children have always played in public and outdoors, often with a degree of freedom and autonomy that twenty-first-century middle-class American parents would regard as reckless and even dangerous, though girls would often be granted less of this opportunity to explore and take risks than boys. Children played in streets and alleyways of cities, as well as in woods and vacant lots. In Victorian and post-Victorian times, children's play was also often accommodated in the home, and play spaces or even a dedicated playroom might be encouraged in twentieth-century parenting and home decorating advice. As the messy, boisterous, unruly family members, children would need durable and rugged interior play areas. The idea of a "rumpus room," as opposed to other names for these spaces, suggests the unbridled, "anything goes" play of the young rather than the more all-ages entertaining suggested by "recreation room." A rumpus room accommodating hobbies like electric trains was often located in a basement, which might be unfinished and rather makeshift, and which might be messy already from its uses for heating, washing, and storage. The shift from coal to cleaner forms of heating fuel early in the twentieth century made residential basements more hospitable to being lived in and used for leisure-time activity. A basement rumpus room could give parents a place to let children play in comfort and safety indoors, keeping them off the street, and perhaps allow for the separation of adults and children into their own spheres within the home. The image of the video game console as a fixture of a carpeted suburban basement, where children and their friends would sprawl on the floor and compete at Atari or Intellivision games, is a product of this history and context.

In mid-century American homes, we also see a migration of play space within the home and its overlap with other kinds of rooms. In his influential 1945 book *Tomorrow's House*, George Nelson advised the designers and architects of what would become the postwar baby boom suburbs to include in the home a "room without a name" suitable for family leisure.[10] In some ways this room's purpose was similar to a basement rumpus room, but the room without a name was to be used by all of the family members rather than only the children, and was to be suitable not only for play but also for entertaining guests. Rather than locating this room in the confined or makeshift space of the basement, it was to be kitchen-adjacent and spacious, the biggest room in the house. Activities to be pursued here were virtually any kind of leisure, including games, media, and dining: "Ping-Pong, bridge, movies, dancing. The children can play there. Or you could cook in the fireplace. Good place for a dinner party too."[11] With a big room like this, the home's living room becomes a smaller and more formal space for adults only. By making this comfortable and versatile big room central to the main floor of the home, the suburban ideal's architectural designs spoke of the virtue of informal family togetherness and integration during times of leisure for family members of different ages and genders: a companionate ideal of pleasure in the company of one's kin. Nelson proposed that the need for such a room was "evidence of a growing desire to provide a framework within which the members of a family will be better equipped to enjoy each other on the basis of mutual respect and affection." The room without a name might "indicate a deep-seated urge to reassert the validity of the family by providing a better design for living."[12] On this basis, Nelson proposed calling this the "family room."

"Family room" was not a novel term, however, having been used at times to describe the sitting or drawing rooms of the Victorian home. The postwar ranch-style house of middle-class suburbs like Levittown were often too modest in size to have both formal and informal living rooms, and either the living room would become a family room (rather than a space in the mode of a Victorian front parlor), or else the family might add another room, DIY style, when time and money permitted.[13] A *Better Homes and Gardens* survey of more than 11,000 Americans in 1946 found that the typical respondent desired a house with a basement for doing laundry and for situating "a multipurpose hobby or recreation room." In Clifford Edward Clark Jr.'s history of the American home, he quotes one survey respondent who desired to combine their living and rec rooms, "but yet have it warm and cozy looking." The respondent described not only its comfortable chairs, but also a pool table easily converted for ping-pong, a refrigerator for drinks, a table for playing cards, a piano, and a fireplace for broiling frankfurters. In Clark's description, this vision of the room "reflected the more informal life-style of younger families. Room design and furnishings had to be tougher to absorb the wear and tear of active family life."[14] In this vision of a house's uses we can also appreciate the idea of the family home as a place for domestic leisure integrating all members of the family together in play.

The widespread interest in making space for play in twentieth-century interiors was a product of a number of social forces, resulting in a conflict in the early years of the century between conceptions of culture and leisure understood particularly in terms of class ideals. As commercial mass media and culture proliferated early in the century—associated in popular conceptions with young, working-class, immigrant, and less educated patrons—progressive and genteel authorities viewed it as a seductive threat to its audiences. An autonomous youth culture of public recreation was seen to be emerging during the 1920s, which guardians of middle-class virtue viewed as a harmful influence on the younger generation in tension with traditional mores. This friction between traditional and modern cultural ideals produced a sense of a crisis in the family, and a need to provide some of the appeals of popular amusements within the space of the family home reconceived as more modern and democratic than in Victorian times. In *Raising Consumers*, Lisa Jacobson argues that this new domestic family ideal made *play* its central value, and play was to be a pastime pursued by children and their parents together.[15] As Elaine Tyler May has shown, progressives of this era frequently attempted to fight mass culture by domesticating it.[16] By making domestic play educational and edifying, it could substitute in the eyes of progressives for the seductive but passive appeals of movies and popular music while also supporting healthy family relations.

Parenting advice of the 1930s, drawing on these discourses of family recreation, recommended establishing a dedicated playroom in the home, which we can see as a point of origin or precursor for the more familiar postwar recreation room. A playroom would be stocked with educational toys and games, as well as books and other media, to combat the supposedly passive spectatorship of popular amusements. Parents, and particularly fathers, were urged to join their children in play, becoming buddies fluent in the slang of their kids and enjoying their companionship. By engaging in this play, the middle-class dad was to "become a boy again," an abiding trope of representations of domestic leisure uniting parents and children familiar from Atari commercials and promotional imagery for many toys and games through the years, including board games, ping-pong tables, and BB guns.[17] Depression-era parents were prompted by *Parents* magazine, which began publication in 1926, to play checkers or billiards with their children and to teach them good taste. This would protect them from less wholesome Jazz Age temptations while also unifying the family in a newly masculinized domestic environment safe from the corrupting influences of the public sphere. "During the 1930s," Jacobson writes, "the family rec room or family playroom gave spatial expression to the hopes that 'the family that plays together stays together.'"[18] In some ways, the recipe for the happy family's playroom was to bring home the popular public attractions of the day and make them safe and morally upright as well as positive forces for domestic cohesion integrating male and female, adult and child participants in domestic living. Movie projectors, dartboards, and pool tables replicated experiences of the cinema and the saloon in the home, but without the threatening connotations of the younger generation pursuing its autonomous culture. We see a similar phenomenon in the domestication of arcade games in the 1970s and '80s. In both instances, the relocation of popular leisure-time pursuits from public to private space functions, in Jacobson's terms, "to reclaim the authority of the family and redefine its mission amid the growing popularity of public commercial amusements."[19]

The family ideal that we associate with the postwar baby boom, prosperity, suburbanization, and the integration of television into the home relied on certain conceptions of a family's common living space. The rooms of the idealized suburban home were designed and arranged to make possible specific activities, including media consumption and gaming. The newly popular split-level plan of the later 1950s would locate the recreation or family room down a flight of stairs from the kitchen, and sometimes in its line of sight. The mother, as commander of the family troops, would benefit from this arrangement by a view from her post in the kitchen down to the play space, making for both autonomous work and play and parental supervision. But this room would be suitable for entertaining, and not be strictly a kids' "rumpus" room where anything goes.[20] The inclusion of a bar in the standard design of a postwar rec room signals its intended adult uses. Along with wood paneling and wall-to-wall carpet, the bar was one of the fixtures of rec rooms. Entertaining, according to advice and trend stories, was moving from formal living rooms to more informal spaces. Articles of the later 1950s distinguished the recreation or family room from the rumpus room in terms of intended users (children only vs. all ages) and indicated that the rec room was becoming the most important feature of a family home, "taking over as the favored spot for family relaxation." The rec room, not the playroom or living room, was the "pivotal feature" in home designs of the "split-level era."[21] Card tables, ping-pong, shuffleboard, and the standard amusements of prewar playrooms were present along with the bar and the newest addition to family leisure: the TV set. In addition to hi-fi stereo equipment, television was the media technology most essential to the rec room, making it a new "focal point of family life" and perhaps taking the place of the movie projector or piano.[22]

Wherever it might be located, the recreation room of the postwar suburban home was not marked off distinctly from other conceptions of domestic leisure space. "Recreation" might emphasize some uses, such as games, amusements, entertainment, and play, but the family room would still incorporate many of these connotations. In publications of the 1970s such as *Popular Mechanics*, readers would be encouraged to finish basements as "rec rooms with a family room look."[23] The accompanying photographs picture a card table, fireplace, upholstered seating, carpet, and wood paneling. A *Popular Science* story from 1975 described a basement rec room as an "extension of living space upstairs."[24] The influence might also flow the opposite way: after the trend of basement rec rooms for family amusement had taken hold, houses were planned with main-story rooms for family recreation, which had migrated upstairs from the subterranean space. "People are less interested in living underground than they used to be," asserted *Changing Times* magazine in 1974.[25] An issue of *Old House Journal* from 1992 portrayed basement playrooms as a short-lived trend as architects soon began to incorporate first-floor play spaces in their plans as a replacement for the less desirable basement rooms.[26] But there is no doubt that a basement rec room, in some ways overlapping in design and ideal use with a main-floor family room, was a standard element and selling point in middle- or upper-middle-class suburban family homes beginning in the 1950s and continuing through the 1970s and '80s. A house without a rec room might be regarded as inadequate to the family's needs of home leisure and entertaining.[27] An essential quality of this space was a function at once of uniting and dividing the family according to its specific needs. It could integrate the whole family together in play and entertainment as in the postwar family ideal, but it could also put distance between parents and the chaos of children's amusement, lessening the pressure of the family's confinement in their house.

A guidebook to home improvement published in 1979 situates the new medium of the video game squarely in this space. After describing the kind of rec room that its reader might build in their basement, DIY style, it offers some examples of activities to pursue there: "everything from the currently popular TV-video contests to the enduring challenge of Monopoly or Scrabble, from pool through darts and ping-pong, from model building to model railroading."[28] Along similar lines, a 1975 issue of *Mechanix Illustrated* published an article on "The New Fun World of Video Games" on a page facing another article, "How to Do Your Basement in Pub Decor."[29] The space represented on the left half of the open pages, a father–son dyad sitting on the shag carpet before an Odyssey plugged into a TV set with *Hockey* on the screen, is virtually continuous with the space on the right half with a staircase, dartboard, wood paneling, and comfortable sofas. These representations of video games position them as children's amusements to enjoy with a playmate dad, but they might also be represented in more adult fashion, as the "pub" theme of the basement rec room would suggest. A 1978 *Playboy* shopping feature titled "Pinball ... and Other Indoor Electronic Sports" offered home electronics such as *Playboy*'s licensed Bally's pinball game for the home and video consoles such as the Odyssey² and Atari VCS as a "*great* way to entertain," encouraging the adult male consumer to practice on his own, the better to impress his friends when they come over.

Along with "TV-video" contests came several other new electronics sold as family or rec room amusements in the later 1970s. Three in particular were closely related to the video game console: videotape decks or VTRs and big-screen projection color TV sets, which together were heavily promoted in the later 1970s as early adopter home-theater technologies; and personal computers, which were a few years later to emerge but, like video games and VTRs, used CRT displays. Along with earlier technologies such as movie and slide projectors and hi-fi stereos, video games, VTRs, and PCs contributed to a widening ensemble of home media, which led to reports of another popular description for a room once "without a name." In many popular press accounts, the recreation or family room of the 1950s era was now becoming a "media room," a term marrying the older meanings of the

Figure 3.3
Mechanix Illustrated, 1975: father–son gameplay on the carpet.

family and rec room with the more modern and cutting-edge connotations of the computer and electronics revolutions.

In many decorating and technology trend stories of the later 1970s and early '80s, the media room was proposed as a place to combine and integrate the various forms of electronic gadgetry in the home, typically centering the space around a large TV set. The typical family probably would not have the means to convert a basement into a media room with screens on three walls, bar seating for five, and a "cockpit" control center, as pictured in a 1976 issue of *American Home*.[30] But representations like these help us form an understanding of the range of meanings for video game technology and its place in the middle- or upper-middle class home. A 1976 *New York* magazine story advertised the media room as "one big custom container for the latest in video equipment," which would make leaving home for leisure unnecessary and even undesirable.[31] *Time* reported that Americans were becoming "chronic stay-at-homes," while seeking entertainment outside of the home was now positioned as an old-fashioned practice of the pre-video age.[32] Descriptions of media rooms often named a video game console as an essential component in an array of electronics to entertain the family, in addition to video decks, film projectors, and hi-fi equipment. "Almost every media room has at least one video game—maybe two or three of the top units. These games take on a whole new perspective when played on a giant screen. Imagine being Pac-Man!"[33] One report noted that Atari games such as *Space Invaders* and *Missile Command* were now commonly being projected on screens measuring seven feet.[34]

The media room would be identified as an outgrowth of both the rec room and the living room, a place for the whole family, but it was also sometimes related to the den, a space carrying more masculinized and affluent connotations.[35] In some accounts, the integration of so much gadgetry into the living space of the home was a problem to be managed, particularly for interior decorators. A *New York Times Magazine* spread in 1981 asked, "How do we make peace with these invaders? How do we integrate them into the home?"[36] A typical solution was to familiarize electronics by adopting the terms of earlier conceptions of domestic leisure, or to anticipate a future of ubiquitous electronic mediation. One description positioned the media room as both a nostalgic throwback and glimpse of the future: "It was a room that would go backwards and forwards at the same time, back to the days when home entertainment brought families together, and ahead to the days when family fun depended totally on electric current."[37] As a place for the latest in electronics equipment, the media room took on some of the male early-adopter qualities of later "man caves" or "tech dens." Sometimes during these years, the newly mediated home was renamed an "electronic cottage."[38] But as a place for families to experience electronic entertainment together, the media room also renewed the spatial significance of the rec room, becoming "the new family gathering spot."[39] In this process, some of the fixtures of the rec room, such as the ping-pong and card tables, gave way to high-tech gadgetry. But as remediations of ping-pong and cards, among other amusements, video games at once replaced and upgraded the older form of family amusement. Early games included not only *Pong* but many variations on card games, particularly blackjack, and many other competitive sports and games such as versions of pinball, hockey, basketball, baseball, and shooting gallery games, in many ways similar to the competitive amusements of the pre-electronic rec room. A media room was a container for not only the latest gadgetry, but for the fantasy of family leisure so much improved by new technology that everyone was happily entertained and contentedly integrated within the domestic sphere.

Video Games as Electronic Family Leisure

Images of console video games in the 1970s and early '80s consistently represented the new product in a particular kind of environment, enjoyed by certain kinds of player groupings. In television and print advertisements, magazine articles, department store Christmas Wish Books, and game company promotional literature such as brochures, instructions manuals, and catalogs, games were often represented in a context of family or rec room play. These images might be spare and suggestive, emphasizing the hardware and the players' enjoyment of it. But they generally would not abstract video games from the setting of the family room, recreation room, or media room. Rather, the *mise en scène* of the video game representation across these various media and formats was quite consistent with the image of the family at leisure in informal communal home spaces. This was likely a product of the audience for these messages being adults and especially women (particularly store catalogs). In many representations, family life is presented in a way that shows off the technology's potential to please all members of the household and bring them together in play.

An image in a 1972 issue of *Popular Science* introducing the Odyssey pictures a man and a boy seated on a carpeted floor before a wooden console television set. The TV is in front of a stone-faced wall, perhaps part of a fireplace. Atop the television set is a floral arrangement.[40] *Mechanix Illustrated* likewise pictured a father–son pair, as mentioned above, in a room with shag carpet, seated on the floor.[41] *Radio Electronics* represented a father–son pair with dad on a sofa and child on the carpet, with the TV set in front of a wooden cabinet and a houseplant to the side.[42] The players are looking not at the TV screen but at each other. A *Popular Electronics* story from 1978 was illustrated with a photo of a man and a woman playing together.[43] Very unusually, these players are dark-skinned, but everything else about this representation is utterly typical: comfortable furniture, wall-to-wall carpet, wood veneer table, houseplants. In some of the images in Coleco and Intellivision catalogs we see no floor, but there is some combination of mixed-gender and mixed-age players, wooden panel or furniture, and greenery. As in many of these representations, one of the group, usually a woman, is looking smilingly at the others rather than at the game, indicating her delight taken in other people's electronic amusement, which brings friends and family together in the space of the home. An illustration in the 1972 Odyssey manual pictures the boy and girl of the family facing the screen, and the father and mother on either side of them, each with a hand on a game controller, facing one another rather than their kids or the screen. As much as they show off the new form of home play, these representations also illustrate the family members, and particularly the mother, taking pleasure in family recreation.

The idea of the video game being a toy for boys in particular is rarely the singular message of these representations, though sons and fathers (as well as mothers) are frequently represented. Rather, the emphasis is on the company of one's kin, as indicated by the photos on the Fairchild Entertainment System console box, which include both adults and children of various ages, including a comically grumpy elderly woman and a baby. The typical image of the video game in use in the 1970s was one of pairs, trios, or quartets of mixed age or gender family members at home.

On the cover of a 1982 Parker Brothers catalog aimed at retailers, we see a tableau integrating such scenes of domestic electronic play within familiar sights and experiences. The cover image represents a fictional Parker Brothers product, "The Christmas Caper Catalog Game," and the booklet's cover is made to appear like a board-game box. Five children of different ages are staged around a living room decorated for the holidays. One stands at the mantle examining an electronic Merlin game with a magnifying glass. A girl sits on a chaise working on the Orb puzzle. Another girl on an upholstered chair uses her magnifying glass to examine a Nerf ball. A boy in a tuxedo in the foreground stands over a Monopoly game. And central to

Figure 3.4
Popular Science, 1972: playing the Odyssey.

Figure 3.5
Radio Electronics, 1975: a parent–child rec room scene.

Figure 3.6
Odyssey manual detail.

this tableau, in the middle ground center, another boy sits on the floor in front of a TV set playing a Parker Brothers Atari game, *Frogger*. As a theatrical staging of these children of wealth fascinated by Parker Brothers toys, this image does not evoke the more middle-class simplicity of the shag carpet rec rooms in the magazines. But it does integrate the newest form of domestic play, the video game (along with another electronic toy, Merlin), within a familiar range of practices and meanings.

Images of family togetherness were standard in representations of home play long before video games emerged as a cultural sensation in the later 1970s. A 1972 Parker Brothers catalog pictured a nuclear family of mother, father, son, and daughter around a board game, under copy promising of Parker Brothers games that "they bring your family together." Postwar advertising often made similar appeals, as in soda, beer, or carpet

Figure 3.7
Parker Brothers catalog, 1982.

ads picturing families and friends in their rec rooms around basement ping-pong tables, or in television set advertisements emphasizing the unity of the family circle. What seems particularly noticeable in early game representations is the social dimension of play and the avoidance of picturing solitary players. Some games, like Odyssey *Tennis*, could only be played in pairs, but many early games could be played by one player against the machine, a setup we rarely see in photographic representations. Rather, these images assiduously mix identities by picturing male and female or child and adult competitors. The frame is often so full of human figures that not all can be actively participating, making some family members into spectators. As the family room was one in which spouses, parents, children, and siblings pleasurably compete with each other, the video game was the latest of the diversions and entertainments to facilitate this familial recreation.

Thus the introduction of electronic games into retail catalogs for stores such as Sears, Montgomery Ward, and J. C. Penney worked to integrate these new amusements within a familiar range of uses and functions. Rather than place such products alongside television sets, stereo components, and other electronic devices, catalogs of the 1970s more often placed video games in the "rec room" pages alongside ping-pong, "rebound" and regular pool, shuffleboard, air hockey, and miniature soccer (foosball) tables, dart boards, and pinball machines. Sometimes they were sold as sporting goods in retail stores, and their high price tags would have made them too expensive to be stocked in a toy store or among the children's toys in a department store. They were marketed at adults rather than kids, who would not have the money for a programmable console, in publications including *Playboy*. In some catalogs the video games were a page away from camping gear like tents and in others they were sold alongside plastic shooting toys and plastic pachinko and pinball games and tabletop electronic football and baseball. They were often advertised using the same kind of pitches as board games from chess and backgammon to Scrabble and Monopoly, as "games the whole family can enjoy." As products purchased most often as holiday gifts, they were aggressively promoted in the fall and early winter and presumably acquired as gifts intended not only for the children but also for the family.

The 1977 Sears Wish Book for the Christmas season opened with video games at the front of the catalog, featuring the newly released Atari VCS branded especially for that retailer as the Sears Video Arcade Cartridge System. The pages picturing these "tele-games" displayed an array of cartridges such as *Blackjack*, *Tank Plus*, *Race*, and *Target Fun*. But this section was labeled in the page corner (by the page numbers) as part of a wider "Family Game Center" theme, which included older, nonprogrammable video games such as Speedway IV, Tank, Motocross Sports, Superpong, Pinball/Breakaway, and Hockey-Tennis II, a ball-and-paddle console. Subsequent pages continue the "Family Game Center" label with more traditional products for the rec room such as darts, chess, checkers, backgammon, pool, ping-pong, air hockey, as well as electronic chess and handheld football games, an array of board games, and electric pinball and pachinko. These pages were filled out not with photographs of rec rooms but with color illustrations, some in a fantastical mode picturing adult athletes and open-wheel racecars alongside electronic sports and driving games. On pages with the more traditional rec room amusements like ping-pong, checkers, and darts, the illustrations were of typical ensembles of middle-class white family members of mixed age and gender engaged excitedly in play, arms raised in triumph or mouths forming contented smiles, with women sometimes observing more than participating.

Catalogs for toy companies bringing video games to market—such as the Parker Brothers "Catalog Game" aimed at retailers rather than consumers—represented similar themes. In Coleco catalogs from 1976 and 1977, that company's Telstar ball-and-paddle console and shooting and driving video games packaged as "Telstar Arcade" are presented in the context of products including CB Radios, pinball and air hockey table games, and doll houses, toy strollers, and toy ovens. Pages representing the video games picture men and women rather than children, in the typical rec room spaces decorated in wood paneling or furniture and houseplants or flowers. Similarly, Mattel catalogs of the early 1980s represented Intellivision alongside Mattel's other electronics products, many of which carry over the types and themes of rec room play into the microchip age. In addition to video game hardware and cartridges, Mattel offered battery-operated electronic versions of baseball, football, racing, chess, and Dungeons & Dragons, using buttons and LED or LCD displays on mostly plastic exteriors, as well as an electronic drum pad and a "Diet Trac" device for weight loss. The photography illustrating these products in use included a number of father–son pairs, and a number of girls and women, though the majority of subjects were male.

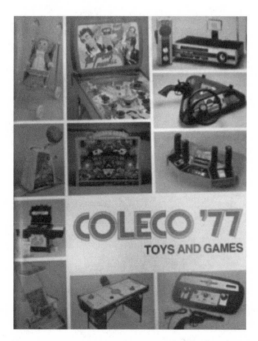

Figure 3.8
Coleco '77 games catalog includes a variety of toys including TV games.

In so many ways, the video game of the 1970s was an outgrowth of earlier forms of rec room amusement, and was experienced as family play in the home, as established before the advent of electronic toys. The clearest and most direct point of continuity between older rec room games and the consoles of the 1970s was the Magnavox Odyssey, a device bridging two eras of home gaming. The original Odyssey game released in 1972 was primitive by comparison even to later ball-and-paddle consoles, and its creator, Ralph Baer, evidently thought that the way it straddled the board game and TV game forms was a failing. Historically, however, this ambiguity in form demonstrates how video games drew upon a tradition of games in the home in establishing their identity, and it reveals continuity as well as change in the history of domestic leisure.

The electronic component of play in Odyssey games was often simple, and the games required additional nonelectronic materials to be fully realized. In most of the original Odyssey games, each of two players could control a rectangle of light by moving it vertically and horizontally, and sometimes the player could put "English" on a ball that bounced between the rectangles, causing it to curve rather than travel straight. Odyssey had neither sound nor color. In some games, the image on screen could be made to move somewhat at random to come to a stop at a point on the display, an effect similar to spinning a wheel or rolling dice. To complete its representation of game spaces, the Odyssey came packaged with translucent overlay sheets that adhered to the glass of the CRT screen by its static electricity charge. Odyssey *Tennis* had a green court overlay, *Hockey* was a white rink, *Football* a green field, *Roulette* a red and black wheel, and so on. These overlays remediated tabletop rec room games such as various baseball and football board and table games, which had been popular for decades in arcades as well as the home. Odyssey also came packaged with a variety of paraphernalia including game chips, cards, dice, play paper money, and game boards. In *Football*, the electronics functioned as one component of a wider ensemble of devices and materials including a cardboard football field. *Football* players would sit across each other with this field between them on the table. Playing Odyssey *Football* also required a paper scoreboard, a roll of frosted tape, a football token, a yardage marker, and six separate decks of cards for different kinds of plays including passing, running, and kicking off. The manual spent six pages describing the process of gameplay, and players would have needed to keep it open while playing at least at first, as is often the case with board games.

Odyssey *Tennis* was essentially *Pong*, and it was self-sufficient even in black and white, though it lacked the onscreen scoring and the blip noise that we usually associate with electronic ball-and-paddle play. But like *Football*, other games were much less coherent as electronic amusements. *Roulette* used the screen as the wheel, but the betting was carried out using paper and plastic pieces. For the geography game *States*, the Odyssey would be used to float the electronic marker somewhat randomly until it would land on a state on the US map overlay. The rest of the game would be carried out using cards and an answer sheet. An *Analogic* game has the player moving the rectangle of light around the screen like the token on a game board. In many of these games, the electronic component is made to fit into a conception of gameplay and materials that has more in common with board or table games than with later video games. And to be meaningful, many of these games relied on the player's prior knowledge and experience of various kinds of play and amusement. Like early cinema, which called on its audience's familiarity with turn-of-the-century genres of storytelling and entertainment to make sense of its representations, early video games drew upon the tropes of pre-electronic arcade and rec room game genres, including casino-style games of chance and adaptations of popular sports. Even *Pong*, as simple and accessible as a game can be, has a meaning based in pre-electronic rec room play.

As its game titles suggest, the Odyssey, like many early consoles, was aimed at families. *Haunted House*, *Cat and Mouse*, and *Simon Says* are juvenile in their cartoonish representations and in their cultural associations. *States* and *Analogic* were meant to be educational. *Shooting Gallery*, a popular title using a rifle controller, was similar to the basement Daisy BB Gun ranges marketed to fathers and sons in the 1960s. Sports games might appeal to boys seeking indoor diversion on rainy days, perhaps in the company of parents, siblings, and friends. That most of these games were meant to be played by two players rather than one indicates the sociable intentions of the producers and advertisers. Like later video game consoles and titles, the original Odyssey appealed to the suburban, middle-class family as a means of bringing them together.

The identity of video games as a medium was a product of this appeal, which continued into the early 1980s particularly in television commercials, on which Atari and Intellivision spent many millions of dollars.[44] Because of their expense, video games were marketed more to adults than to children, and their value and interest had to be made clear to parents. In one TV commercial from 1980, a white middle-class family of four sits in their family room playing video games in the evening, a stack of cartridge boxes next to the Atari on the coffee table. The son and daughter are on the floor, while their parents sit on the sofa. The father is nearer to the TV set and sometimes is playing. The mother, never represented holding the joystick, is by a telephone on the end table that never stops ringing. A voice-over intones sardonically, "This commercial is based on a true story." While the TV set bleeps and blips and the kids and dad engage with *Space Invaders*, the mom picks up the phone as a succession of babysitters call to offer their services, which she declines again and again. "After a family bought an Atari video game," the announcer explains, "they had no trouble getting babysitters ... Everybody enjoys Atari because Atari has so many different games to enjoy." Parents in this comical representation were so delighted by Atari that they preferred to stay home and play in the company of their children, overcoming any desire to have adult-only leisure away from home.

Other Atari commercials of this vintage showed boys and girls playing Atari games including *Berserk* and *Pac-Man* with their grandparents, and one portrayed a succession of family members (though not the mother) enjoying their favorite games, with a final shot of their pet dog manipulating an Atari joystick. A common image in the video game advertising of this period is of the family crowded around the player holding the Atari joystick, with some members on either side of him or her on a sofa and others standing over their shoulders or sitting by their feet. Video game producers and marketers, along with magazines and retail catalogs, strove to insert Atari and similar amusements into a familiar context of domestic play enjoyed by children, parents, and even grandparents all together, unified by their delight in this new amusement. If the reputation of video games developed in a different direction, it was against the background of this rhetoric of a cohesive, companionate family ideal.

Boy Culture and Escape from Domesticity

In spite of the efforts of manufacturers and retailers to market early video games as family amusements for all, the culture of video gaming quickly developed in a different direction, one in tension with the meanings of the more inclusive discourses of these catalogs and advertisements. Video games did not become boy culture merely as a consequence of being a favorite pastime of male players. They also drew in many ways upon a history and context of masculinist socialization and representation. To call this "boy culture" is not to deny the presence of female players or their significance in a history that often erases them. But it is to assert a masculinized identity emerging during this time, defining the medium by practices and meanings that are hardly neutral in terms of gender, but rather participate in reproducing patriarchal relations of difference and power. Boy culture is not meant to refer simply to the culture of boys and only boys, but to culture understood by association with normative youthful and masculine gender identities, defined largely in opposition to the feminine.[45] The dynamics by which video games became boy culture played out in public, in the arcade, but also in the space of the home, the private, feminized sphere wherein video games struggled to establish a place as the province of boys and men.

All evidence suggests that during their first decade, video games in the home were becoming masculinized in terms of the identities and relations of players. Social science literature of the early 1980s found that video games in the home were more often purchased and played by fathers than mothers. Fathers and sons were often found to play together, but many mothers (half in one study) had never played them at all six months after acquiring a console.[46] Fathers, however, were found often to compete at first, only to abandon video games when their children bested them. Boys and girls both played in significant numbers, though not equally: boys played for longer average times than girls.[47] And in families with sons, Edna Mitchell writes, "possession by the boys was considered appropriate; and sisters had to request permission for access to the games."[48] With programmable games, peers could share and trade cartridges, but this developed as a male culture, and girls were not found to trade carts or take them over to each other's houses.[49] Mitchell's study, published in 1985, described how video games, upon initial purchase, would bring families together as playing cards and board games had previously, but that after the novelty wore off they became the children's amusement, and mainly the playthings of boys. A conference on video games held at Harvard University in 1983, at which a number of psychologists and other social scientists presented research, included discussions of the preference of male more than female players for this new form of amusement, a difference with sources in gendered play dynamics that apply in many instances of games of chance, physical skill, and strategy.[50]

A commercial for Atari's *Defender* cartridge suggests the gender dynamics at play in the culture of early video games. For most of the thirty seconds, a man plays the violent spaceship shooting game with a woman sitting by his side on the sofa. At the end she finally gets to play, and the kicker is his expression of surprise and disappointment: "You did better than me!" This makes the address of the ad more expansive than just a young male player, but it also underscores the gendering of video games as masculine, as the interest and prowess of the woman functions as more of a punch line than a pitch.

By the early 1980s, the video game industry was aware that its primary market was males aged eight to eighteen, though it also continued to appeal to families.[51] This appeal might rely on a distinction between the arcade and home, with the latter promoted as a safer alternative for children. Spots in a "Dear Atari Anonymous" TV campaign are narrated by a woman, the mother of a family in which Atari has taken over everyone's leisure time. In a comical tone, she complains about how video games have transformed the family and made her children and spouse into addicts. But this complaint was gently satirical, and conveyed not only new media anxieties but also a sense of security that home video games would promise. One of their appeals was the advantage they offered as alternatives to arcade games. In a spot for *Berserk*, a boy's grandmother wants to take him to play at the arcade, but he delights her when he says they can do it at home. With home versions of the arcade games, younger children would be safe from the perceived threats of the world outside, and parents would be relieved of worrying about kids squandering pocketsful of quarters in the coin-operated machines.

A 1981 video game industry survey found that 90 percent of American arcade players were male, and 80 percent teenagers, and the shady reputation of the arcade, along with the associations of teenage boys with unruliness, were factors in keeping electronic play within the sanctuary of domestic space.[52] Many parents preferred to supervise their younger children's leisure-time activities, or at least to keep it within the safety of the home. While one ideal of play was companionate leisure, games were very often played by a solitary child, most often a male one.[53] By 1982, video game consoles were in 17 percent of American households. In homes with teenage boys, however, the percentage would be much higher.[54] The most desired game titles, moreover, were home console versions of arcade sensations like *Space Invaders*, *Missile Command*, *Asteroids*, and *Pac-Man*, many of which featured aggressive and violent Cold War representations of space battle and heroic defense against overwhelming enemy forces. The image of good-natured play in family-directed marketing discourses was not the same as the typical everyday experience of games.

In appeals to the family market, video gameplay was represented as a sociable activity, but games were frequently enjoyed in solitude, as children's toys often are. As Brian Sutton-Smith argues, video games are merely a particularly engrossing form of children's amusement that can be pursued in solitude for seemingly unlimited amounts of time, an ideal toy for functioning to free parents for a time from the burdens of care and companionship.[55] Console games might have substituted for TV watching when little of interest to children was on the air, or for other forms of mechanical play. Most games, such as cards, board games, and table games like ping-pong, require two or more players, but video games could be enjoyed by one person alone. Sutton-Smith's explanation for the appeals and functions of video games connects them not so much with sociable, companionate ideals of the rec room as with children's toys such as blocks, dolls, and trains used to pass the time while parents take respite from attentive care or do housework. Such solitary amusements, Sutton-Smith proposes, also function to prepare young people for any kind of future work in which "individual and solitary concentration on the task at hand is a requirement."[56]

Whether experienced in solitude or in the company of peers, siblings, or parents, early video games drew in a number of ways on a history of children's play in which boys and girls have been socialized into distinct and opposed worlds of fantasy and activity. As Steven Mintz describes in his history of American childhood, since the nineteenth century, middle-class children's play has been divided by gender as boys and girls were assumed to differ in many fundamental ways, both physical and psychological, and as they were preparing for adult life in societies of marked gender-role divisions. Girls were encouraged to see themselves in their mothers, making virtues of "self-sacrifice and service," while boys' identities were defined by negation of femininity, stressing "aggression and daring" and condemning girlish boys as sissies.[57] The nineteenth-century American home was feminized, and boyhood was "defined in opposition to the confinement, dependence, and restraint of the domestic realm."[58] The play of girls and boys was a product of the social relations prevailing in industrial society, with feminine pursuits training girls for care of home and child, and masculine pursuits stressing adventure, courage, competition, battle, and physical strength. Boy culture of the nineteenth-century United States inculcated autonomy and independence in male children, contrasting the freedom of masculinity in contrast to the domestic confinement of femininity.[59]

In the twentieth century, typical experiences of childhood changed in some ways, with more schooling for children of both sexes and expansion of cities and suburbs. The gender roles of children's play continued, however, as commercial culture drew on the Victorian era's meanings and practices, and as play became an escape into fantasy worlds far from the realities of classrooms and families dominated by adult authorities. In books, movies, and television for children and in consumer products tying in with them, girls and boys were often divided and their cultures of play maintained. Boys were sold Buck Rogers pistols, plastic soldiers, and Davy Crockett hats and rifles. Among the most successful toy products for girls in the twentieth century was the Barbie doll, which modeled adult feminine appearance and pursuits. According to Mintz, children's literature and children's toys provided a "simulacrum of reality for increasingly structured lives," a characterization that fits well with the fantastical representations of early video games. It also functioned to socialize young people into prescribed gender identities.[60]

The meanings in children's play can be regarded in relation to patterns of childhood development, but they can also be read as symptoms of wider social and political realities. The Cold War gave rise to a gendered children's culture in which boys in particular were socialized to act out symbolic struggles of good against evil, emphasizing a heroic and active conception of masculinity. The simplistic morality of movies and television for boys, particularly in the western genre so popular in the 1950s and '60s, was mirrored by the development of masculine children's culture emphasizing aggressive, competitive, physical activity. Being considered tough was the highest honor for young male children. Boy culture in this period was at least in its ideal form a free-ranging outdoor culture, and by contrast the postwar girl culture of slumber parties was more confined to indoor space. Iconic 1970s video games such as *Space Invaders* and *Missile Command*, with their science-fiction confrontations in defense of humanity against fearsome foes, reproduce the Manichean morality and aggressive confrontation of Cold War boy culture. Most of the very popular games of this era married the *us vs. them* moral universe of westerns with sci-fi scenarios, as in *Star Wars*, the most successful cultural phenomenon of the time. Some were more traditionally masculinist in their generic representations, however, including the Activision title *Chopper Command*, a game in which the player is in control of a helicopter engaged in military combat. In a commercial for this title, a drill sergeant yells at his troops and the voice-over addresses the player: "If you've got the guts, we've got the game." The sergeant in his tent plays *Chopper Command*, clutching the joystick like a weapon and using the fire button to control the game's helicopter as it engages in a dogfight.

The genres of early games drew heavily on boy culture scenarios and iconographies of long standing, including many kinds of competitive sports and shooting games with settings in theaters of war, the American West, jungles and deserts, and especially the space opera backdrops of post-*2001* sci-fi. Graphics were often quite abstract in the first and second generations of consoles, but box, cartridge, catalog, and advertising imagery would flesh out the settings and characters and more generally the connotations of a game's representation. The illustrations in this imagery fixed a meaning for sometimes vague or inscrutable onscreen representations. These images located experiences of video games in a recognizable world of popular culture narratives and imagery familiar from books, magazines, comics, movies, and television. Atari's shootout game *Outlaw* represents stick-figure cowboys firing across a simple cactus plant. The cartridge box art shows paintings of a covered wagon drawn by a team of galloping steeds, a pair of bearded gunslingers with their revolvers drawn, a smattering of gleaming gold coins over the bottom of the image, and an orange sun setting over Monument Valley in the background. *Combat*, another Atari title, offers twenty-seven games around two main concepts: fighting with either tanks or planes. The onscreen image represents an overhead view of a pair of tanks that maneuver around a maze to shoot at one another, and an overhead view of planes against a cloud field in the flying version. Its cartridge art pictures military tanks in a line along a desert landscape of dust and fire, as well as a warship, biplanes, fighter jets in formation, and missiles, all in dynamic action. And Atari's *Maze Craze*, true to its name, represents one figure pursuing another through an intricate labyrinth. The cartridge box fixes the meaning of this chase as "cops and robbers," picturing a police officer wielding a nightstick hounding a robber around a warehouse, both figures shown in active mid-stride with excited facial expressions.

Some games were so abstract as to allow for much interpretive embellishment, though the various options would all derive from the imagery and ideals of masculine play. *Super Breakout*, an update of the earlier Atari arcade and console hit *Breakout*, is a kind of sideways *Pong* in which a paddle at the bottom of the screen moves right and left to bounce a ball against lines of colored bricks on the top of the screen, which vanish when hit. *Breakout* had been narrativized in commercial representations using a jailbreak scenario, with the bricks of the game standing for the wall of a prison cell. But *Super Breakout* was narrativized in its cartridge box art with a representation of astronauts in space suits playing a racquet sport against a backdrop of planets and stars, combining imagery of two boy culture standbys: space adventure and competitive athletics. An audio book set released in 1982, *The Story of Atari Breakout*, combines a story in text and illustrations with an audio recording to be used in combination with the

(a)

(b)

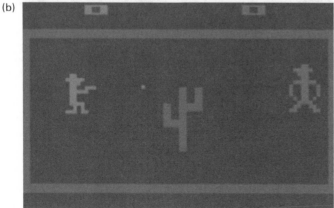

Figure 3.9
Atari *Outlaw* cartridge box and game.

(a)

(b)

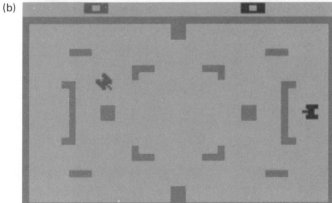

Figure 3.10
Atari *Combat* cartridge box and game.

(a)

(b)
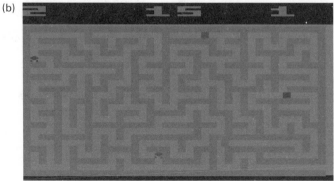

Figure 3.11
Atari *Maze Craze* cartridge box and game.

Figure 3.12
Story of Atari Breakout, audio book set cover, 1982.

book, telling yet another version of a *Breakout* scenario. In this account, the game represents a space shuttle transporting valuable ore from a moon of Jupiter to a space center called New California orbiting Venus. In this narrative, *Breakout* is about firing missiles against the colored layers of a force field obstructing the path of the shuttle. Upon the astronaut's success at penetrating the force field, the book narrates: "He was doing it, breaking through, he had won! What a triumph for a son of earth, for Captain John Stewart Chang!"[61]

Historically, boy culture meant a free-ranging, outdoor experience of exploration and autonomous play in streets and alleyways, woods and lots. It gave independent young men the freedom to learn the values of courage, bravery, stoicism, and daring, far from the watchful eyes of mothers or teachers. In the era of video games' emergence, however, middle-class American children's play increasingly moved indoors under the supervision of adults fearful for children's safety. Beginning in the 1970s, American media has provoked strong moral concern over the welfare of children and the dangers of unsupervised, autonomous children's play.

These fears have centered on sexual and violent dangers posed by strangers in public places, such as abusive adults, reckless drivers, and kidnappers. As a consequence of worries over children being left alone, Mintz argues, young people since the 1970s have rarely been allowed the freedom to explore on their own in "unstructured, unsupervised free play."[62] The children of such fearful societies have often been confined indoors to amuse themselves under adult supervision, and have spent more of their childhoods alone rather than among peers, to be amused by electronic media. Video games emerged just as middle-class families curtailed the freedom of their sons to venture off on their own, and those sons substituted the mediated adventures of Atari games for earlier experiences of outdoor exploration and discovery. Video games participated in the "islanding" of childhood, as kids were confined in the security of the domestic sphere, protected from the perceived threats of a morally corrupt society.[63]

Thus video games came to function as "virtual play spaces," in Henry Jenkins's apt formulation, reproducing indoors many of the qualities of outdoor boy culture.[64] The independent, tough and aggressive, competitive play of boy culture, with its emphasis on speed, strength, and endurance, continued in the forms of representation and gameplay present in many early games, with their spaceships, explorers, racecars, athletes, tanks, cops and robbers, and western gunslingers, their joysticks and fire buttons. A measure of the freedom lost as a product of childhood's islanding was regained in the imagined fantasy environments suggested by the rather abstract representations of 1970s and '80s TV games. Like children's literature and popular culture of earlier Cold War forms, video games of this period represented life-and-death struggles and elite athletic competitions. Mastering such games was no small feat, and won players admiration and status among their peers. But this way of understanding early games, these associations with boy culture and gendered play, sit uneasily with the meanings of the rec room and its association with companionate family leisure and a gentler brand of competitive play.

As a consequence, representations of gameplay in some discourses opposed the integration and harmony of many catalog and advertising images. Rather than finding their place in the comfort of family gathered around a new electronic hearth, video game discourses aiming at young male players in particular rejected the familial and the domestic. In one Atari ad for *Space Invaders*, a family plays inside while outside, at night, their house is comically besieged by the descending alien lines of the game. This image of the domestic sphere under assault by murderous

Figure 3.13
Atari commercial: *Space Invaders* descending on the family home.

invading hordes is a fitting emblem for the counterdiscourse of the more masculinist rhetoric in Atari game ads, fighting the image of the home as a comforting sanctuary and threatening it, however humorously, with disruptive violence.

Such discourses also positioned gaming as an escape from domestic space into boy culture fantasies, leaving the home behind. Similar to the later PlayStation ads Bernadette Flynn describes representing games as an "electronic portal to a virtual exterior," outside the feminized domestic realm, representations of video games in earlier discourses also offer imagery of departure or escape.[65] Often the body of the male player is removed from the scene of the home or becomes enveloped in the world of the game. The images of the male player addressed by these ads is often of a soul lost to electronic microworlds, totally absorbed in the representation, leaving everyday reality behind to enter completely into the identity of whatever agent is implied or pictured in the game (pilot, astronaut, racer, fighter, etc.). The boy or man is transformed by video games into someone else, somewhere else. What he most clearly and centrally leaves behind is the home in which the Atari console was located. *Time* magazine referred to video games in 1976 as "Jocktronics," and described the games in terms of masculine identification: you are an athlete, a race car driver, a blackjack high roller. Several cartridge advertisements of the late 1970s and especially the early '80s literalize this departure and transformation, picturing destruction of interior spaces and physical changes in the player.

Atari cartridge ads, particularly in the early '80s, often addressed a young male player directly with voice-over narration in the second person. An ad for the game *E.T.* (a notorious failure) pictures a player resembling the movie's central character Elliot, implying that you can become the young hero through electronic play, and instructs the boy to have his parents hook up the console to the TV set. Another commercial is scored to a new wave song with vocals made to sound electronically generated. The lyrics put the player into the game: "You're a starship captain in an asteroid field, blast away your lasers or put up your shield." These lines are sung over cinematic images of a spacecraft followed by a screenshot of *Asteroids*. It continues with a screenshot of *Missile Command*: "And a missile commander defending your city, if you're not breaking up they'll show no pity." Then it cuts to *Space Invaders*. "Invaders won't stop us from the sky they drop." Only at the end of the spot do we see children playing the game, as parents look on, smiling. The combination of these three sci-fi shooting games unifies the masculinized experience of Atari play around the same themes of battle and defense as the boy is assaulted and urged to fight back. Often the player is assumed to identify with the game, even to the point of losing a sense of self located in external reality. A *Tunnel Runner* Atari ad shows a teenage male running through a computer graphics maze, and the voice-over describes how, when you play, "you don't look down on the maze, you're in it." The first shot is of a TV set on which an image of the player is superimposed on the game, showing the male figure transported into diegetic space.

In some representations, ordinary life is shown undergoing transformation through game experiences, making for dramatic consequences. The Activision game *StarMaster* was advertised in a minute-long spot representing a young man returning home, dropping his keys, and petting his cat. When he begins to play the space-themed game, the lights in the room extinguish and the player seems surprised; beams flash in his face as a deep-voiced male narrates, "Fight and fight again, retreat to refuel, battle and be blasted right out of your senses." The player relaxes by playing and becoming totally absorbed in the game, achieving "flow." Then a computer graphics effect connects the player's head to the screen in bright rays of light. Extreme close-ups of the player's face alternate with the game screen making for a sense of very close connection and total absorption in the space adventure and battle depicted; his eyes widen with amazement and intensity.

Figure 3.14
Activision *StarMaster* commercial: the player is being brought into the game.

The ad for a 20th Century Fox Atari game from 1982, *Beany Bopper*, went deeper into the absorption theme. Johnny is a boy playing Atari alone in his room to the scoring of new wave synth music, with a lyric that begins, "Winning takes your total concentration." A female sing-songy voice calls, "Johnny, telephone, it's Susie … Johnny, Daddy is here with your new puppy … anybody seen Johnny?" The camera pushes closer and closer on Johnny's face, framing him tightly in intense concentration, coming to an extreme close-up on his eye. The song lyric goes, "The only way to win it is to really get into it." At the end of the ad, Johnny's parents enter the room to find him absent, and they peer into the television set and the game contained within it. This is where their son has gone, and the spot finally cuts from their gaze into the virtual realm, which has claimed their boy, to the cartridge box, the product promising such experiences of absorption and escape to the sons of the American family. This representation of gaming relays parental anxieties about electronic play, but also the possibility for video games to transport their male users, separate them from their socialization among parents, siblings, and peers, and take them elsewhere, escaping from domesticity, heterosexual courtship rituals, and familial bonds. Through video games, they might leave all of these behind for a world of masculine adventure.

Some images of escape also were images of destruction. A 1982 ad for the 20th Century Fox Atari game *Mega Force* was typical of a number of ads made for products on this label, using a contemporary rock score and a male player of late teen or early adult age. The young man enters a bedroom with posters on the walls and a guitar on the bed and begins to play *Mega Force*, straddling a chair. Immediately his head is shown covered in a motorcycle helmet and quickly it cuts to the game screen. But within a moment we leave the game imagery in favor of cinematic representations of motorcycles, tanks, explosions, and smoke. When we return to the home, the bedroom wall has been blasted away to reveal a landscape beyond of the fantasy imagery of the game, with wind making the lamp and clothes hanging from wall hooks swing back and forth. As in many ads for early '80s games addressing male players specifically, the *mise en scène* shows a violation of domestic interiors, invasion from the game narrative, and breakdown of the reality/game distinction.

A minute-long MTV-style spot for Atari's 1982 cartridge *Centipede* crystallizes a number of these appeals, showing the male player leaving the family or living room of the middle-class home to enter a different space, another story world, upsetting the comforting image of home. It begins with a young man sitting on the family room floor playing the game, in which you fire from the bottom of the screen, as in *Space Invaders*, at creatures descending down from the top. Rather than aliens reminiscent of *The War of the Worlds*, the enemies in *Centipede*—an arcade hit at the time—are creepy-cute bugs. After a few moments of absorbed play, an insectoid appendage emerges out of the TV set and pulls the player inside. We then see a tabloid newspaper headline: "Centipedes Invade!" A pastiche of images follows of various film genres: horror, documentary, silent melodrama, all of them depicting something frightening or unsettling or panicked. Generically this is a mashup, but conceptually it shows the influence of killer bug films of the 1970s such as *Bug*, *The Swarm*, and *The Giant Spider Invasion*. The new wave theme, a sound-alike of the *Phantom of the Opera* title song, is set to a chorus of "Centipede!" sung in a portentous, breathy whisper. In one scenario, a starlet is asleep beside a human-size bug, and when waking to discover who is in her bed, she bolts upright and screams. The ad cuts to another image of a screaming woman's face, this one in black and white. Quickly we see another headline, "Marines Battle Insects!" followed by newsreel footage of soldiers, warships, and bombs. When the commercial cuts back to a screenshot of the game, the rhythmic fire of the guns seamlessly blends into the fire of the game's weapon against the descending Atari creatures. Finally, after displaying the cartridge box, the young man returns to his family room from inside the TV set, but by playing *Centipede* he has undergone metamorphosis into a bug the size of a man and calls out, "Help!"

Like the "Atari Anonymous" and "Johnny" commercials, this spot gently satirizes games for ruining the lives of their players, but it also delivers a pitch to players eager to be seduced by the power of the new medium. As in many of these representations, we see rather little imagery of the game itself, which is pedestrian by comparison with the lively pastiche of film styles. The rhetoric appeals more on the basis of associations players might make with the game's scenario, here with cinematic rather than computer iconography. But most centrally, the experience of a game like *Centipede* is shown to afford an escape from everyday reality into a movielike realm, an experience so profound as to transport the young male player away from home and family, and to transform him into something else.

The idea of video games effecting a transformation of the male player's identity was one of the most common tropes of visual representations of the new medium in the late 1970s and early '80s. These Atari cartridge commercials often figured such a change as confusion between domestic and game space, but the notion of the player becoming someone else or being transported someplace else was also visible in many representations less invested in negating domesticity explicitly. Illustrations in the magazine *Electronic Games*, which began publication in 1982, often represented the fantasy elements of video gameplay. One illustration from an early issue shows a pair of boys sitting facing each other, their hands gripping joysticks. They are pictured seated on a white cloud against a sky blue background, and below the cloud their joystick cables lead into a pair of football linemen, who they control and effectively become through the fantasy of play. The boys above are light-complexioned, while the adult athletes are dark-skinned, adding an element of racial fantasy to the ideal of play as departure from everyday identity.

Similar to the "into the game" representations of the commercials, *Time* magazine's cover from January 1982 represents a male player's body entering into a fantasy realm. Its headline blares "GRONK! FLASH ZAP! Video Games Are Blitzing the World," and the illustration below is of a lone male figure exchanging gunfire with a flying saucer in the screen of an arcade cabinet, picking up on the same appeal as in the Atari commercials of a player entering game space (see figure 1.1). Like the *Electronic Games* illustration, this image promises that for the male player, video games would be not merely diverting and exciting, but transporting. The real-world self would slip into a new identity, far from the scenarios of home and family

Figure 3.15
Electronic Games, winter 1982: a boy fantasy of play as escape.

used most often to position video games as the latest iteration of companionate rec room leisure.

To arrive at these images of masculine identity play, we have traveled far from the cheerful rhetoric conveying the unity and inclusiveness of video games, in which players of different ages and genders come together in the home even as the boy is most centrally addressed. Early game imagery was not unified in its messages about space and identity. In some ways it inserted video games into an already established scenario of family play in the suburban rec room, but in other ways it extended another play tradition located more often outside than in. The contradictory appeals of early game promotion, at once harmonizing the family in the home and offering the boy a virtual escape hatch, not only reveal the challenges of domesticating a new medium of electronic leisure, making familiar its uses and meanings. They also express the contradictions of American family life and childhood development during the later Cold War years, as boys were confronted with the electronic mediation of their adventuresome culture in the confinement of the domestic sphere.

4 Video Games as Computers, Computers as Toys

Many buyers of video game machines might not realize it, but what they are really buying are the electronic innards of a home computer.
—*Newsweek*, 1982[1]

There's a saying: "You can tell the man from the boy by the price of his toy." Whoever coined that phrase must have been a computer buff.
—*Radio-Electronics*, 1982[2]

In a scene in *National Lampoon's Vacation*, the American film comedy of 1983, a father enters his suburban family room where his son and daughter are using a home computer connected to a television set. He tells his children to stop playing video games and put down their joysticks so that he can show them the "TripTik" he has programmed using the family computer to plan their car travel from Chicago to California. Cut to a map of the United States on the TV screen: a car is animated departing from Chicago trailed by a dotted line marking its route to Wally World. But right away the children pick their joysticks back up. The son controls a Pac-Man-like figure who pursues the family car while gobbling the dots marking the route. Then a space alien sprite controlled by the daughter appears, firing at the Pac-Man and saving the family station wagon from the fate of a video game ghost devoured along the way to a high score.

This representation conveys generational tension between parents and children and a comical assertion of paternal authority, but it also speaks of the place of computers and video games in American society at the moment when *Vacation* was made. Home computers at this time were caught between adult and child uses, between seriousness and fun. The father's novel use of a computer to plan a highway route fits into a history of masculine tinkering with technology, with home computers emerging in the late 1970s and early '80s as a popular hobbyist pastime and a new

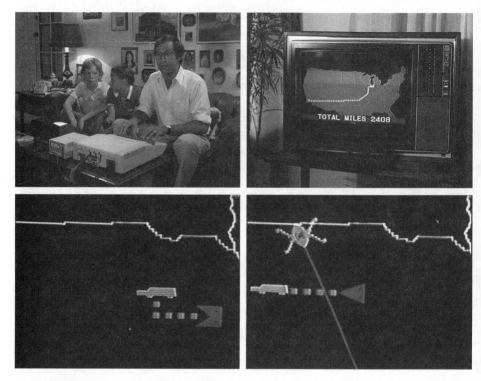

Figure 4.1
In an early scene in *Vacation* (1983), the use of a home computer to plan a trip is hijacked by the children's video games.

productive tool with an array of functions. The children's use of the new machine to play video games also fits with typical conceptions of computers in the home. While often imagined as useful in virtually infinite ways, early home computers were undoubtedly most often used not for supposedly productive purposes like accounting or data management, but for playing the same kinds of games available in arcades and on video game consoles. The instability of the computer's identity, caught between play and work and between meeting child and adult needs, was central to its emerging cultural status. Other scenes in American comedies of the later 1970s and early '80s portray advanced electronics in similar ways. A heart monitor display is used to play *Pong* in the Mel Brooks comedy *Silent Movie* (1976), while in the comical spoof *Airplane!* (1980), an air traffic control radar screen is used to play *Basketball*, an Atari console game. Representations like these find humor in the confusion around the appropriate functions of screens, in the incongruity of important tools being used for pleasure and diversion rather than to meet more legitimate needs.

Video games for the home were already familiar by the time computers for the home emerged, and the two were tightly connected and overlapping, helping to define each other's uses and their respective places in popular imagination. Some of the key figures in the emergence of home computers, particularly Steve Jobs and Steve Wozniak, had some background in video games, and Jobs had been employed by Atari before he founded Apple. But computer games were not new in the 1970s, or even in the 1960s, when *Spacewar!* was programmed for an MIT PDP-1 minicomputer (small and user-friendly for a computer of its time) connected to a CRT output. Computers were programmed to play games before they were typically used with graphical displays. Throughout the mainframe and minicomputer era, many of those with access to computers used them "off the clock" for nonwork purposes, typically programming them to play text games or *Spacewar!*, depending on the kinds of input/output available. Games were often important in computing pre-1970s in several ways, as programmers solved problems and experimented through game programs, explored the potentials of their tools, and employed games as demonstrations to show off what a computer could do. (*Spacewar!*, for instance, was demonstrated to the public at MIT soon after it was created, and the PDP-1's producer, Digital Equipment Corporation, eventually also used the game in demonstrations.)[3] People, particularly young people, learned to use computers by programming and playing games.

The introduction of microprocessors on silicon chips in the early 1970s was one factor pushing the proliferation of electronics devices and their increasing use in everyday life for many purposes. But another factor was a widespread desire of people to use computers, which had been seen as powerful institutional tools of organization and control, and as futuristic artificial brains.[4] When those outside of institutional settings accessed computers for the first time in the 1970s and '80s, their encounter was often defined and understood in terms of playing with the tools. Whether they used computers and other devices containing microchips, such as calculators and handheld games, for work or fun, their access to this new technology was playful in the sense that people were newly able to explore the potential of objects that had been previously out of reach, off limits. While some middle-class children might have accessed computers in the 1960s and '70s through schools and time-sharing access, and thousands of minicomputers were sold at this time for scientific and industrial uses, a popular perception of computers held that the machines were supposed to be for official work within corporate or state bureaucracies. The history of the personal computer is one of ordinary people gaining access to advanced technologies in order to play with them, and quite often to play video games.[5] The transformation of computers from an emblem of control and depersonalization to a means of individual liberation was particularly apparent in the countercultural rhetoric of techno-utopian authors like Ted Nelson and Stewart Brand predicting a future of computing as a form of creative exploration.[6]

At the same time as personal computers (often "home computers" in the parlance of the 1970s) became familiar, video game consoles became more computer-like. Apple's breakthrough model, the Apple II, was released around the same time as Atari's Video Computer System, a console in which the game cartridge contained a microchip programmed with a particular game. Several of the video game consoles of the later 1970s, such as the Magnavox Odyssey[2] and Bally Professional Arcade, either came with a QWERTY keyboard built in or had a peripheral keyboard available for separate purchase to "turn the video game console into a computer." At the same time, handheld games such as Parker Brothers' Merlin and Milton Bradley's Simon were marketed as toys to be used in competition between a human player and a computer. From microwave ovens and calculators to hi-fi tape decks and clock radios, microchips were entering into people's everyday experiences, including their leisure-time play. Video games had been represented more in relation to television in the earlier years of the 1970s, before this more widespread computerization of everyday life. By the end of the decade they were more typically considered as computer objects, while home computers themselves became associated with TV screens serving as their video display outputs.

The most popular mass-market home computers of the later 1970s and early '80s, such as the Apple II, Radio Shack's Tandy TRS-80, Atari's 400 and 800, Texas Instruments' TI-99 4/a, and the Commodore PET, VIC-20 and C64, were thus caught between two conceptions and ideals of their functions and uses. For many consumers, the less expensive models (those excluding Apple) were often seen as video game consoles with the added value and legitimacy of home computers, which could be programmed to do more than just play games. Atari's computers had a particularly blurred identity, carrying the most famous corporate identity in video games and accepting game cartridges just like the VCS/2600. Unlike earlier home computers such as the Altair and Sol, aimed at electronics hobbyists and subscribers to titles like *Popular Electronics*, the PCs of the late 1970s and early '80s were marketed at home electronics consumers who might have little interest or expertise in workshop tinkering, kit building, and experimentation with gadgets and tools like soldering irons. They were "turnkey" products, easy to use even to those without a background in electronics.[7] Many of the early home computers were advertised and promoted as video game devices with extra functionality, essentially as gateways to computing trading on the familiarity of video games but adding the seriousness, usefulness, adult interest, and excitement of computing to justify a higher cost in relation to game consoles. They were also sold, similarly to consoles, by appealing to the whole family. But rather than unifying the family in play, as game ads promised, the home computer would appeal more as a something-for-everyone product. The kids would use them for homework and games, while the parents would use them for managing family finances, organizing recipes, and even telecommuting.

These first widely used home computers in North America—particularly Apples, Ataris, TRS-80s, and Commodores—were caught between two conceptions of value. Early video games were an application of an emerging new form of computing if not computers themselves, but they were also playthings for young people. The tension between these conceptions can be given as a set of related oppositions:

adult/child
tool/toy
labor/leisure
serious/fun
important/frivolous
productive/not productive
legitimate/not legitimate

These pairs, moving from concrete to abstract and from descriptive to evaluative, represent the conflicted place of both video games and computers in the culture into which they were emerging. The early home computer was not simply or straightforwardly a serious tool of work, but was caught between that identity and another as a plaything, an outlet for individual creativity and pleasure. This flexibility of meaning produced unstable ideas about computers and their uses, which blurred the boundaries between the opposed terms.[8] While acknowledging that video games often turn out to be the most popular way to use a computer, discourses of the period often treat this as an inconvenient, almost shameful fact, suggesting that the appropriate use of the home computer was to serve more important and legitimate functions, meeting the needs of adults as well as children, and educating as much as amusing their younger users. At the same time, the appeal of video games helped sell these computers even as the useful functions of the microcomputer legitimated its expense and presence in the home. Calling something a computer or describing it as "computerized" or "computer-powered" was a selling point for a variety of products in these years, particularly games and toys. Computers had cachet.

They were also seen to be central to the development and education of young people, who were to mature into an information and technology age that would demand familiarity with computers in their everyday lives, particularly in the more respectable occupations. Whether using educational software or "just" playing, children growing up with computers and electronic toys were learning to use the tools of the future and becoming fluent in their operation. As hobbyist, tinkerer, and early-adopter electronics technology, computers already were emerging as a particularly masculine cultural form. While many women worked in computing fields, the worlds of the academic, military, and business computer were often male-dominated, with differing levels of status accorded to male and female roles within the workplace.[9] The computer as a home technology built on these gendered associations, and the more youthful and future-oriented aspects of the technology's cultural identity were also tinged with these unequal gender relations. A culture of home computing emerged in these years, overlapping the culture of video gaming, which established and maintained a hegemony of masculinity in the realm of advanced consumer electronics. Boys and girls, as well as men and women, were all users of early home computers. But the ideal user as represented in dominant discourses of the time such as magazine and trade publications and print and TV advertisements was a man or boy, or perhaps a father and son together. Female computer users were not excluded altogether, but they were marginalized in a number of ways as a masculine home-computing culture became entrenched. Boys and men and their play with computers and games provided the basis for the explosion of PCs into popular consciousness and into the texture of everyday life, to the point that *Time* magazine named the computer its "Machine of the Year" on the cover of its January 3, 1983, issue. The illustration on the cover, naturally, pictured an adult male user. (When the front cover is unfolded, it also pictures a woman accompanying him with her own computer, her secondary status affirmed by being hidden from view on the cover as displayed on newsstands.)

Figure 4.2
Time "Machine of the Year" cover, 1983.

By the later 1970s, video games (as well as other amusements, such as Merlin and Simon) *were* computers, while computers were often seen as game-playing devices, and more generally as objects with which ordinary people might play. These were changes in the identity of both video games *and* computers, which informed the cultural ideals inscribed in both. As these two categories blurred into one another, the computer with which games were so closely associated was an object with flexible meanings. While the countercultural roots of the personal computer would promote play as individual exploration, often the marketing and promotion discourses through which many consumers encountered home computers emphasized their utility for white-collar work and conventional education, as well as their function as a video game console. Even if they were unlikely to be used for supposedly productive, adult purposes, their image as a tool of the office drained away the more techno-utopian fantasies of countercultural visionaries by making computers seem appropriate for purposeful information labor. The tensions and contradictions between these meanings—between the use of computers for work and for play, and between the useful play and playful work of many experiences of computers and games—were central to the emerging identity of both home computers and games. Games and play were essential to the development of PCs as consumer products, and to the establishment of their place in popular imagination. Computing was likewise crucial to the identity of video games as it was being formed.

Computer Games to Video Games

Games have often figured into the history of computers and not just as diversions; 1970s video games were not new as devices using computers for play. Many of the central concepts applied to games played on home computers were holdovers from the era of mainframes and minicomputers used not in the home but in universities, government agencies, and institutions with substantial calculating and record-keeping needs, such as public utilities and airlines. The world of computer games before video games and computers became consumer products established discourses through which the games and platforms of the 1970s and '80s would be understood.

Early in the history of digital computing, machines were programmed to play familiar games such as tic-tac-toe, checkers, chess, and Nim. Computer programmers of the 1950s might have undertaken this for fun, but programming a game might also be a way to explore the potential of the machine and demonstrate its functions, while also showing off the computer's abilities to nonexperts.[10] Seeing a computer play a game like checkers on television (which occurred in North America in the 1950s) might demystify the potent power of the machine for the general public.[11] At the same time, it could also reinforce a sense of its almost magical abilities. But even putting aside the image of the almighty computer in popular representations, some game programs proved to be important developments in both the history of computing and the history of games and play.

Claude Shannon's famous essay "Programming a Computer for Playing Chess" appeared in 1950 and framed the idea of a computer playing games in terms of use-value and learning: "Although perhaps of no practical importance, the question is of theoretical interest, and it is hoped that a satisfactory solution of this problem will act as a wedge in attacking other problems of a similar nature and of greater significance."[12] He listed a number of productive possibilities for computer chess, proposing that accomplishing the goal of teaching a machine to play the game would lead the way to more impressive accomplishments, such as computerized logical deduction and strategic decision making. Arthur Samuel's checkers-playing computer program, first demonstrated in 1954 but a product of years of work, was regarded as a significant milestone in the field that would become known as artificial intelligence. When Samuel's IBM 701 computer beat a human opponent in 1956, the theme of the machine prevailing in its battle against man made this seem like a momentous occasion, and IBM's stock was said to have increased in value 15 percent as a consequence. Observers were so impressed that a myth took hold that computers had "solved" checkers, and that no human opponent could beat the computer.[13]

Shannon's chess-playing program was just an idea in 1950, but a few years later, machines were competing with human beings at a number of games. A 1958 *Scientific American* story, "Computer v. Chess Player," describes the challenge of programming an IBM 704 computer to play chess as an "intriguing" one, as it requires a computer programmed not merely to compute but to think.[14] Several research groups were pursuing this challenge in the mid-1950s, including one made up of the article's authors, employees of IBM. While not able to consider every possible move, the 1958 chess-playing computer described in *Scientific American* could be programmed to avoid some human blunders, and to capitalize on its flesh-and-bones opponent's mistakes. That IBM found this project to be a worthwhile use of a computer designed to make engineering and scientific calculations speaks to the productive uses of games in computer programming in these years. Computer chess continued to be an important area of research in computer science for decades, and annual computer chess championships were held into the 1990s.

Notably, however, discussions of game-playing computers in the 1950s and '60s typically engage with questions of legitimacy. The British computer Nimrod, a machine for playing the game Nim (in which players take turns removing sticks from a set of rows until only one stick remains), was demonstrated publicly in 1951, accompanied by a booklet explaining and justifying its existence. This text makes clear that that the Nimrod has been constructed not merely for fun, but "to demonstrate the principles of automatic digital computers."[15] The authors anticipate skeptical responses, insisting on the validity of their enterprise. "It may appear that, in trying to make machines play games, we are wasting our time. This is not true as the theory of games is extremely complex and a machine that can play a complex game can also be programmed to carry out very complex practical

problems. It is interesting to note for example that the computation necessary to play Nim is very similar to that required to examine the economies
of a country in which neither a state of monopoly nor of free trade exists."
In other words, while this object may seem frivolous, its uses are actually
serious and practical and will fulfill important functions for our future
society.

While the 1952 tic-tac-toe game OXO (known in the UK, where it was
made, as Noughts and Crosses) used a CRT display, the more typical input/
output for computerized games in the 1950s and '60s would be switches,
buttons, lights, punched tape, or teletype. Chess and checkers games
required entering positions and moves and waiting for the machine to output its response. Photographic representations of play typically showed a
man seated at a regular game board set up on a computer. But the move
toward playing computer games with interactive video outputs did not necessarily change the status of computer games as productive rather than
merely diverting efforts.

Spacewar! is often regarded as the first computer game and video game
not so much because it was the first game to use a computer or a video display (it was neither), but because it was a game that was created for a computer with a video display that could only be played with a graphical
computer system.[16] Similar to computer chess, checkers, tic-tac-toe, and
Nim, however, *Spacewar!* was not merely an effort at using advanced technology for fun. It was, rather, a "display hack": an exercise in showing what
might be possible with a computer outputting to a visual display, an innovation in programming. It showed not only the magic of a computer giving
you control over a spaceship moving around a screen and firing at enemies,
but also the potential for real-time interactive computing. At a time when a
standard use of computers involved submitting stacks of punch cards for
batch processing by an operator and then waiting for the delivery of output, a computer terminal like the one used for *Spacewar!* allowed users to
give and receive inputs and outputs instantly and feel the full power of the
machine under their control. It was novel in allowing you essentially to talk
to a computer and have it answer back to you at once. An account of the
game in an MIT newspaper, *Decuscope*, in 1962, notes both its competitive
space adventure aspect, but also its demonstration of the "real-time capa-
bilities of the PDP-1." Steve Russell, the student programmer who created
Spacewar!, tells the paper that the program's "most important feature ... is
that one can simulate a reasonably complicated physical system and actually see what's going on."[17]

While *Spacewar!* might have been created as an effort to solve a problem and explore the potential of technology, it quickly became an object of fascination and a way to play with computers absent the more instrumental utility of making a clever hack or demonstrating a type of program. The *Spacewar!* code was copied and improved upon, circulating among computer users around the United States who competed in *Spacewar!* tournaments and worked on modifications to the game. In the 1960s and '70s, versions of *Spacewar!* were played wherever people had access to mainframe or minicomputers connected to CRT displays, often at night or when supervisors weren't paying attention.[18] A number of text-based amusements were also popular, including *Star Trek*, *Lunar Lander*, and *Adventure*. The culture of computing, limited to institutions with the means and needs for these expensive machines, incorporated play with computers as a somewhat illicit diversion. Playing games with computers in the era of the mainframe and minicomputer was sometimes represented in popular discourse as a waste of precious processor time. In a 1974 report on *The Game of Life*, a mathematical amusement that could be played with or without a computer or display, *Time* magazine contrasts the legitimate use of a technological resource, a "precious commodity," with the frivolity of spending it on a diversion. "Computer specialists everywhere have developed such a mania for Life that millions of dollars in illicit computer time may have already been wasted by the game's growing number of addicts."[19] By participating in the popular text game *Star Trek*, in which the player assumes the role of the captain of the starship *Enterprise*, computer users were engaging in an "underground phenomenon," according to the historical account of Paul Freiberger and Michael Swaine, "hidden in the company or university computer's recesses, played surreptitiously when the boss wasn't looking."[20]

Mainframe and minicomputer games of the 1960s and '70s, the precursors of the home and arcade video and computer games of the 1970s and '80s, were seen to have contradictory values, depending on how they were used. These values would endure into the era of the programmable home console and personal computer. If games were programmed as a way of learning the machine's potential or working out some programming problem such as artificial intelligence (chess), integrating graphics and real-time processing (*Spacewar!*), or managing a database (*Adventure*), games might be valued highly and considered worthwhile.[21] The 1968 book *Playing Games with Computers* aimed to teach young people computer programming through games, which could be copied from the book into a computer program in FORTRAN.[22] Programming computers to run casino games like

Blackjack and Keno and "mathematical recreations" like pentominoes and magic squares would be useful in instructing newcomers to programming in the input of instructions in a computer language to solve well-defined and already understood problems. This made them useful in computer science classes.[23] The People's Computer Company of Menlo Park, California, a grassroots group that started up neighborhood computer centers for education and recreation, published a book of games in 1975 in collaboration with Hewlett-Packard called *What to Do after You Hit Return, or P.C.C.'s First Book of Computer Games*. *What to Do* promoted computer games as creative resources for learning computing but also for education more broadly, developing cognitive skills while also having fun.[24] As a countercultural group, it saw no contradiction between play and work, and preferred to see them interwoven in the experience of computing. Whether countercultural or more mainstream, though, programming computers to play and demonstrate games was often regarded as productive ways of using computer technology. If computer games were merely for play, however, having no function for learning or exploring the computer's potential, they were often seen as a waste—as an illegitimate use of the most advanced and valuable modern technology. Individuals found value in merely playing with computers; institutions did not.

Personal Computers as Video Playthings

The PC has its origins not only in the development of a new form of computer processor, the silicon microchip, but also in a social rather than technological cause. There was a widespread growing desire during the 1970s to use computers. This desire was to make accessible a technology regarded at this time both as promising and modern and potentially quite powerful and liberating, but also as a tool and symbol of institutional control and depersonalization under the military-industrial complex. The effort to make a computer for individual use was in part a power-to-the-people movement, with roots in the 1960s counterculture.[25] The *Whole Earth Catalog*, a counterculture bible, was a central point of reference for many of those pushing for personal computers. But equally strong was the force of individual hobbyists, the tinkering young men in Silicon Valley garages who had access to computers at their jobs at companies like Hewlett-Packard, but who wanted computers of their own, to use on their own time and for their own purposes.

What people were actually going to do with these computers was not always seen as that important in the 1970s. The goal was to own the machine and get it "up," that is, functioning.[26] Frequently the inevitable question of "What can you do with a computer?" was answered with something like "What can't you do?" followed by a laundry list of bright ideas, usually untried, along with an obvious answer: play games.[27] A *Time* magazine story described home computers intended not only for TV games but to brew tea and roast turkey, water the lawn, lock the doors, and control stereos, televisions, lights, and drapes, though it did not make very clear whether some of these functions were merely to be imagined.[28] A 1977 *Esquire* story described a computer fair in Trenton, New Jersey, where male hobbyists sold parts off the tailgates of their station wagons. The author likened these men to 1950s high-fidelity enthusiasts, "hard-pressed to decide just what to do with their electronic power once they have it." The point might often be to tinker, assemble, solve problems, and impress peers with technical achievements, and the goal of using a computer might be subordinate to the pleasures of these pursuits. "None of them *needed* a home computer any more than a mechanically inclined teen-ager *needs* a 400 HP coupe with cheater clicks. The idea was to get a piece of the action."[29] Ultimately the best justification for owning a computer was that it would help you learn to use a computer; more instrumental functions would come later. This has been called the "self-referential" stage in computing, when technology itself was the center of interest rather than using it for some external purpose.[30] Thomas Streeter argues in *The Net Effect*, his book on the romanticism of computer culture and its importance in the development of the Internet, that the origins of the personal computer are most centrally in a deep interest in playing with computers to satisfy individual passions and to learn the ways of the machine. This would free computers from their use as instruments of command and control, and put them in the hands of people rather than institutions. Computers could then serve personal and creative ends rather than merely bureaucratic and institutional imperatives.[31]

It is no exaggeration to say that this desire was to change the popular conception, as well as the actual uses, of computers from instrumental to recreational. The motto of the Bay Area People's Computer Company was "If work is to become play, then tools must become toys." *Byte* magazine, which began to publish in 1975 as the unofficial organ of the PC movement, proclaimed computers "the world's greatest toy!" on the cover of its first issue in 1975. The idea of the computer as a plaything, a hobbyist's diversion, predates the existence of consumer commodities like the Apple II, which gave its owner the ability to play with a computer; Apples and the rest of the consumer market computers of the '70s arose to satisfy interests expressed by these discourses.

The first commercially successful and widely publicized and popular microcomputer available for home use was the MITS Altair 8800. On the outside the Altair appeared as a metal box with lights and switches, hardly resembling what we now picture as a personal computer. It had no keyboard or screen for input/output and no way of storing data, though one could add peripherals. After appearing on the January 1975 cover of *Popular Electronics*, the Altair enjoyed brisk mail-order sales to hobbyists, both as a do-it-yourself kit and as an assembled unit.[32] The typical customer would have had experience with computers and electronics from work or school, and a desire to tinker and play with one on his own. The user could program the Altair (rather laboriously) in a version of BASIC using switches, with output in blinking lights. The Altair also contained slots for adding memory and input/output devices. The pleasure of this computer was that the machine was fully under your control and ownership, and it would respond to your input with its output. The technology itself was the object of interest, rather than achieving some external outcome by using the machine as your tool.[33] The Altair, often regarded as the first step toward the PC revolution, was a microcomputer toy for electronics enthusiasts. Around 1977, thousands of white-collar consumers purchased microcomputers as business expenses, though they had little business software available to them and used the machines mostly for their own amusement.[34]

It was not the ultimate ambition of computer hobbyists merely to flip switches and make a panel of lights flicker on and off. A loftier goal of the 1970s computer culture was to bring television screens to life and control images on them using computers, much as Russell and his MIT cohort of hackers had done with *Spacewar!* Although the Altair does not much resemble what we think of as a personal computer, it is not too far a stretch to say that a PC is basically an Altair with a typewriter keyboard, television screen, and a memory storage device such as a tape or disc drive, and Altairs could be configured to include these things. A combination of typewriter keyboard and TV screen would allow hobbyists to use personal computers to play the games familiar from workplaces and university labs, like *Spacewar!*, *Adventure*, *Star Trek*, and *Life*. An important step in the direction of such a machine was the "TV Typewriter" of Don Lancaster, featured on the September 1973 cover of *Radio-Electronics*. This was not a computer per se, but an input/output ensemble consisting of a modified typewriter and television set.[35] If decades later the ambition of typing letters and seeing them instantly appear on a television set might seem underwhelming, to the 1970s hobbyist this ideal of real-time computing was a potentially notable achievement.[36] And the novelty value of the TV Typewriter persisted into the later years of the decade. A number of video game consoles of the later 1970s, such as the Magnavox Odyssey[2], included programs that allowed the player to do no more than type alphanumeric messages on a television screen.

But merely making characters appear on a screen was not the most imaginative or exciting dream of the computer culture. One of this culture's most vivid and utopian expressions belonged to Ted Nelson, author of *Computer Lib/Dream Machines*, a rich and poetic text first published in

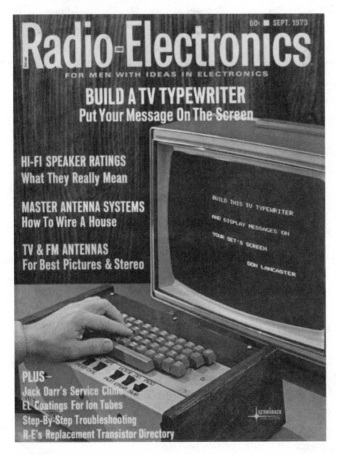

Figure 4.3
"TV Typewriter" cover of *Radio Electronics*, 1973.

1974 (and frequently reprinted for the decade to follow) and modeled after the *Whole Earth Catalog*.[37] Among other things, this book is known as the source where Nelson coined the term "hypertext," but its ambitions are much broader than that well-known fact suggests. *Computer Lib/Dream Machines* functions both as a primer on all things computer, and as an imaginative program for the future. The book has two front covers, two halves, and two related but distinct topics. The *Computer Lib* side explains computers, makes the case that they are "marvelous and wonderful," and argues for their liberation from institutions for use by ordinary people through slogans like "EVERYBODY SHOULD UNDERSTAND COMPUT-ERS" and "COMPUTERS BELONG TO ALL MANKIND."[38] The image on the cover of a raised fist recalls not just revolutionary movements like black power but also women's liberation. (Robin Morgan's second-wave feminist anthology *Sisterhood Is Powerful*, published in 1970, has a similar cover.) Nelson compares computers to bicycles, automobiles, and cameras, as technologies that can be made to serve their owners' individual needs. Along the way he mentions games such as *Spacewar!* and *The Game of Life*, but mainly as illustrations of some of the ways individuals have put computers to use. *Computer Lib* is one of the most important examples of the rhetoric of the computer culture of the 1970s anticipating and agitating for personal computers.

On the cover of the book's flipside, *Dream Machines*, the illustration functions not so much to instigate a revolution in the present as it does to express a fantasy for the future. A young man in a Superman cape flies with his finger pointing the way toward a rectangle of white light, a screen. The subtitle is "New Freedoms through Computer Screens—a Minority Report." Nelson proclaims this side of the book to be "exhilarating and inspirational," and what he means to inspire is a particular excitement about computers as outlets for a specifically visual, moving-image form of individual creative expression. Computers outputting to TV-like screens will allow users to "write, think, and show."[39] In particular, outputting graphics to cathode-ray tube displays, which he excitedly references as "lightning in a bottle," will be a way to engage in creative and personal ways with computers.[40] "IF COMPUTERS ARE THE WAVE OF THE FUTURE," Nelson prophesies, "DISPLAYS ARE THE SURFBOARDS."[41] These displays would be not merely graphical like a movie but responsive to the user in real time, which was the mode of computer usage called "interactive." Nelson breathlessly asserts: "If you have not seen interactive computer display, you have not lived."[42] The most exciting prospect of personal computing, to Nelson, is an interactive experience in which the user manipulates graphical outputs. The ideal future use of computers for people rather than institutions would be a device that can "MAKE PICTURES and show you stuff and change what it shows depending on what *you* do." Nelson warned his reader: "If this idea doesn't turn you on, the rest of this book is probably not for you."[43]

Another intellectual turned on by computers with interactive displays was Stewart Brand, the editor of the *Whole Earth Catalog* and a central protagonist in the story of the influence of counterculture on cyberculture.[44] Brand published an essay in *Rolling Stone* magazine in December 1972, and, along with *Computer Lib/Dream Machines*, "SPACEWAR: Fanatic Life and Symbolic Death among the Computer Bums," became an essential document for understanding the personal computer's genesis in techno-utopian post-1960s ideals.[45] The two writings have in common a special interest in visual and interactive computing, but unlike Nelson, Brand in this instance is as much New Journalism reporter as guru, and he makes a computer game into his central example leading the way forward into the new, individualistic conception of a computer. Visiting computer researchers and technicians who compete in the First Intergalactic Spacewar Olympics at Stanford's Artificial Intelligence Laboratory (SAIL), Brand describes their play and the science-fiction elements of the game (spaceships, thrust buttons, enemy torpedoes, "5-4-3-2-1," "Ohhhhhh NO! You killed me, Tovar!"). These competitors are at once "heads" and "hackers," kids staying up all night in "outlaw country." To Brand, computers are comparable to LSD in their potential cultural impact. They are not represented as wasting time or resources except in a positive, antiestablishment way. The hackers' frenzied and joyous play shows what a computer can do and can be when applied to the pursuit of individual passions. *Spacewar!* is what will happen when computers are under the power of people rather than institutions. Brand quotes Alan Kay, the Xerox computer scientist working at the Palo Alto Research Center, to the effect that *Spacewar!* "blossoms spontaneously wherever there is a graphics display connected to a computer." It is merely natural that computers with TV-screen displays will be used for a pleasurable, absorbing game in which spaceships move around under the user's real-time control and fire at one another.

Like Nelson, then, Brand construes fun as the ultimate benefit of technology. The goal is to appropriate a tool of work and control as a personal plaything, and to reveal play and creativity as the machine's most ideal use. "Until computers come to the people," Brand argues, "we will have no real idea of their most natural functions." *Spacewar!* was a potently countercultural example of alternative computing because it was the work of creative, visionary hackers rather than official planners, and was no one's idea of a computer's official, legitimate uses. It functioned for Brand as an auspicious sign that the hackers were taking over from the planners, that they were "cultural revolutionaries."[46] Games would strip the military-industrial and bureaucratic functions out of computing and by doing so make computers serve the needs of humanity, as understood according to the New Communalist vision of the *Whole Earth Catalog*. Brand called *Spacewar!* a "flawless crystal ball" representing a future of thoroughly interactive, individualized computing, and proclaimed: "*Spacewar* serves Earthpeace." Putting aside the more outlandish utopianism of such aphoristic expressions, we can see in Brand and Nelson two clear expressions of what the historian of technology Paul Ceruzzi calls the "mental model" of personal computing: the full power of the machine being put in the hands of its user.[47] Video games provided a key example of this model, and *Spacewar!* in particular demonstrated a dynamic of machine–user interaction in which personal rather than institutional goals are at stake, and pleasure rather than utility is the desired outcome.

Games like *Spacewar!* also offered an example of a computer whose graphical video output is its strongest appeal. In addition to mainframe, minicomputer, and microcomputer games and innovations such as the TV Typewriter, a number of other efforts were made in the mid-1970s to animate computer-made images on a television display. TV Dazzler was a color graphics card released by Cromemco in 1976 for use with a CRT display outputting from a computer using an S-100 bus, such as an Altair. It could be used to play a version of *Game of Life* but could also generate kaleidoscopic imagery or alphanumeric characters. A screen image of *Game of Life* in color, played using the TV Dazzler, appeared on the cover of *Byte* in June 1976. The Kaleidoscope program for the TV Dazzler was described in *Esquire* as "something nearly as pretty as a tropical fish tank."[48] Around the same time, in 1976, Atari released its Atari Video Music visual synthesizer, a colorful music visualizer that connected to both hi-fi and TV set, creating similarly psychedelic imagery. Other S-100 bus computers with video displays, such as the Sol and IMSAI, came on the market in this period. A Sol computer programmed with a colorful shooting game, *Target*, was demonstrated on Tom Snyder's network late-night TV talk show *Tomorrow* in 1976, amazing the host, and presumably many in his audience.[49] In Steven Levy's account, this TV demo functioned to familiarize the general public with microcomputers. "*Target* was perfect for showing Tom Snyder and a television audience a new way to look at those monsters shrouded in evil, computers."[50] The program itself, like *Spacewar!*, might be regarded as a "display hack" worthwhile for showing potential as much as utility.

Efforts to familiarize computers through video, from *Spacewar!* to the TV Typewriter and TV Dazzler, speak to the centrality of the screen to the meanings of a computer as understood in everyday life. It was not adequate for the microcomputer to be under the individual's control. To appeal to individuals, it also had to do something interesting and fun, and that thing was most often visually engaging games. When Steve Wozniak dreamed of a computer he would make for himself, according to his autobiography, he had in mind the kind of machine that would play the games he had programmed and played at work at Hewlett-Packard and Atari, such as *Star Trek* and *Breakout*. When he first saw *Pong*, Wozniak found it interesting not so much as a game but as a way of manipulating an image on a TV screen, a key appeal of the computers he subsequently created.[51] The color graphics, sound, and game paddle peripherals included with the Apple II are evidence of his desire. You could do so many things with an Apple II, including playing and perhaps even programming *Breakout*. Unlike earlier computers that resembled airplane cockpit controls with their lights and switches, the computers that succeeded as consumer products to be purchased for the home used the familiar input/output of typewriter keyboard and television set. Apple II was made to be unthreatening and inviting: Steve Jobs took inspiration for its curved molded plastic casing from the Cuisinart food processors he had seen at Macy's. These consumer products, which often used a television set as an output, were really not so different from earlier devices used to play video games in the home.

Fun with Microchips

The image of the computer changed in the 1970s in a number of ways. If we simply consider popular representations of computers, we find shifts and openings, marking developments in popular conceptions of technology. In the hugely successful and influential 1968 film *2001: A Space Odyssey*, the computer HAL is represented as a human-like thinking and feeling being with a man's voice, but visually he is personified mainly as a red light, as in a shot/reverse-shot sequence with the astronaut Dave, his interlocutor. This interaction between human and machine occurs through speech, in colloquial American English. This is of course an idea of a future computer decades along in history, but it is notable that the computer imagined by the filmmakers little resembles either the computers of the 1960s or their successors only a few years away.[52] Their representation of a personified machine endowed with something like consciousness and powers of thought and feeling informed later ways of thinking about computers, but also failed to offer a very clear vision of future technology. Typical representations of computers around the time of the film's release pictured the standard institutional array of large machines with buttons, switches, and lights. Computers used punch cards or teletype for input, and large reels of tape and rolls of paper for data storage and output. In popular imagination, the thinking and feeling machine might still be years off, and computers were regarded mainly as tools to make computations and store data, less as technologies with communicative uses or functions, or applications for everyday use by individuals. During the 1970s, however, advances in semiconductor technology coupled with demand for consumer electronics made for new ideas and images of computers.

Consider two covers of *Time* magazine, one representing "The Computer in Society" in 1965, and the other "The Computer Society" in 1978.[53] In the 1965 image, the machinery is overwhelming and multifaceted, functioning as one big brain controlling an array of functions. The computer itself is represented as reels of tape, buttons and lights, punch cards, and teletype all issuing from a cerebrum much like a human being's. The machine is an unsettling combination of animate and inanimate, organic and inorganic, with its arms extending from a metal body, gigantic in scale next to the human beings working beneath it. The tape reels are above a mouth dispensing punch cards to a stack held over the head of a woman collecting them on a platter. This image prompts questions along the lines of: is the machine working for the people, or is it the other way around?

The 1978 cover is quite different. Although the hybrid man-machine motif carries over, the tone is much more cheerful. The image of the computer is more varied and human. C3PO, the humanoid robot from the previous year's blockbuster sci-fi film *Star Wars*, presides over a group portrait in which each human figure has a computer-related object in place of a head, suggesting that we are becoming computers, or that computers are helping us to think and perceive the world around us. These computer-related objects are a mixture of old and new, more and less familiar. We see a punch card and magnetic tape data reels to one side of C3PO. But there are also more novel consumer artifacts for individual use: a digital wristwatch, a pocket calculator, a silicon microchip, and a personal computer terminal pairing a QWERTY keyboard and a video display. The presence of the cheerful robot from a movie aimed at children, familiar as much from toys as from the cinema, introduces a more familiar and comforting tone. The subtle change from "computer in society" to "computer society" also

(a)   (b)

Figure 4.4
Time covers: "The Computer in Society" (1965) and "The Computer Society" (1978).

indicates a change from thinking of technology as tools people within institutions use to thinking of the tools being woven into the everyday life of ordinary people and their social fabric. The watch and calculator were both microchip devices for one's personal use, small enough to fit in a user's palm or on their wrist. And while it might seem that play is absent from this representation, calculators in particular were objects often regarded in the 1970s as high-tech toys for people who had never previously had the opportunity to use a computerized technology for their own purposes and pleasures. For instance, a volume entitled *The Calculating Book* was published in 1976, with the subtitle *Fun and Games with Your Pocket Calculator.*[54]

While the computer of ten years past *2001* was markedly different from HAL, two future-minded conceptions of the 1960s persisted. Computers continued to be represented as brains—as thinking machines—and they also continued to be personified by being given a name and identity, and by linguistic conventions that referenced them as individuated, person-like beings.[55] The handheld games Merlin and Simon were both described on their packaging and in promotional discourses as computers. Like many other examples of computerized toys and games of the later '70s, many with "computer" in their names and package descriptions (e.g., Computer Perfection; Comp IV; Zodiac, the Astrology Computer; Electronic Detective, the Computerized Who-Done-It Game; and Code Name: Sector, the Computer Game of Submarine Pursuit, and many chess computers), these gadgets were represented as playmates for the user rather than merely as toys containing electronic components. The electronic competitor, an identity that video games would often adopt, was one of the central ways in which microchips entered the everyday lives of Americans in the later 1970s.

The television commercials advertising Simon and Merlin, products of two of the biggest toy and game firms at the time, Milton Bradley and Parker Brothers, made the computer component central to the pitch, and gave the product agency and individual identity like HAL, as well as HAL's particular combination of friendliness and inhumanity. A commercial pitching Simon, a pattern-matching toy with four large plastic buttons and a speaker emitting electronic tones, conveys this through the voice-over announcer's script: "Simon's a computer, Simon has a brain. You either do what Simon says, or else go down the drain. Simon is a master, he tells you what to do. But you can master Simon if you follow every clue." The description of a microchip as a brain was familiar from many representations of consumer electronics in the 1970s, such as a line of calculators called the Bowmar Brains. One ad for its products pictured a grinning college graduate dressed in a gown and mortarboard, holding up the product under copy that reads: "I got my looks from Mom, my drive from Dad, and my brain from Aunt Tilly." By making the microchip into the brain of a playmate, however, toy and game companies were adding intentional states to the more computational connotations of that usage. Now that the brain is playing a memory game or tic-tac-toe against you, all sorts of qualities, such as feelings and desires, can be attributed to a machine.

A commercial for Merlin, a handheld electronic toy resembling a telephone handset with a grid of light-up buttons and a speaker, emphasized the more humanizing aspects of the brain metaphor, going beyond mere calculation. A pop jingle "Where's Merlin?" accompanies shots of the members of a nuclear family each using Merlin for a different purpose. It begins, "Where's Merlin? Where did it go?" as a boy searches for the electronic toy under his bed. The singer answers, "Janie's got it playing tic-tac-toe," as a girl holds Merlin. As the boy continues to search for the product, we see his father playing Blackjack 13, his brother playing Magic Square, and his mother playing Music Machine; Mom finally surrenders the playmate to her son before the ending product shot. Halfway through the ad, the announcer explains, "Merlin is a computer with personality, plays six different games, talks in twenty sounds." Another commercial from the same time described Merlin as "a remarkably intelligent computer" and informed the audience, "you compete with him." The meanings of microchip technology in these spots for Simon and Merlin combine the power and novelty of the computer with the friendly competition of home play among family members, familiarizing and domesticating microprocessors while also selling the image of them as powerful thinking machines with the agency and identity conveyed by a personal pronoun. The meanings of home computers as friendly high-tech toys and thinking machines were continuous, informing ideas about handheld electronics but also ideas of the more powerful and multipurpose machines that would ultimately be known as home computers.

Ascribing agency to the computer was a rhetorical move made by many kinds of writing on these new forms of technology, not just the ones with cute first names. A 1978 *Popular Mechanics* first-person essay about acquiring and learning to use a TRS-80 home computer is largely an account of learning to use the technology through play, and most of what the author, Stephen Walton, does with his TRS-80 is play games like *Blackjack*, which was included with the hardware. While he does attempt to learn to use the computer for purposes presented as more legitimate than games, such as household management, he confesses that, after six months with the technology, "mostly, my computer plays games with me."[56] He has copies of David Ahl's *101 BASIC Computer Games* and *What to Do after You Hit Return* (Ahl's book was to be found near many a home computer in the later 1970s), which he uses to program games, and professes greater pleasure from programming these games (the more legitimate use) than merely playing them.[57] He describes this in terms that give the machine an identity, observing that the computer never gets tired, noting that it has a flair for drawing pictures on the TV screen, and speaking of programming as bending the machine to his will. Often promotional materials for computerized amusements in this period represented their use as this kind of struggle of man versus machine, with the challenge presented as a feat of outsmarting an artificial mind potentially superior to the human. An advertisement for Magnavox's Odyssey^2 game console, with its QWERTY keyboard built in, promoted it with the slogan: "The excitement of a game. The mind of a computer." The language of the ad copy made clear that the keyboard in particular was the component giving the user "access to the mind behind the games." In various ways, the attribution of mental powers to computer playthings carried on the themes of dangerously powerful technology from *2001*, while also selling the novelty and futurism of technological companionship and competition in the electronic age. This helped to establish the worth of computer playthings as advanced tools of thinking, as toys of rational thought for sharpening human minds. The emphasis on computerization, on the competition of human players with artificial minds, with sophisticated electronic competitors, gave legitimacy and importance to experiences of playing with toys. These experiences augured new ways of living in a technologically advanced society, giving consumers the opportunity to hasten to bring the future into their everyday lives. But compared to programming and managing data, the use of computers for mere play was sometimes treated with skepticism and doubt in the later 1970s and early '80s, as potentially frivolous or wasteful. In the discourses of marketing and in the press accounts introducing the home computer to readers, computers could be useful or fun, but rarely just fun.

An Electronic Swiss Army Knife

When the home computer emerged as a consumer electronics product around 1977, it was introduced to the public as an object with multiple overlapping functions, meanings, and identities. Unlike the utopian rhetoric of Stewart Brand and Ted Nelson, many applications of the home computer promoted in advertising and marketing discourses emphasized the use of the computer as an instrument of white-collar information labor serving institutions as much as individuals, and productivity as much as creativity. Not all of the multiple uses of the home computer were to be equal, and not all users were expected to share the same interests and expectations. The more legitimate and adult uses of the home computer for work, business, and household management were combined with the more juvenile uses of the computer for video games, and the serious and fun uses might be in competition with one another. As the product cost hundreds of dollars, the buyer would typically be a middle-class or wealthy adult, and the potential range of users might include children and their parents. Uses might include both workplace and household management tasks along with educational programs, games, and even just becoming familiar with a computer. Its image as a combination of a typewriter and a television set, which is how it was often described in stories about the first home computers, serves as an apt emblem of the new technology's

Figure 4.5
Magnavox Odyssey^2 advertisement: "Mind of a Computer" signified by a QWERTY keyboard.

place in popular imagination. It was at once a tool for traditional managerial work and a source of diversion and amusement. While the typewriter helped sell the product as useful and important, most computer users young and old would likely play games on it, and the computer was often promoted as a more sophisticated version of the game console. Gaming familiarized the computer, while the keyboard and the potential to program made it seem more worthwhile to own a device that still might mainly be used for fun.

The imagery and narratives used to promote the home computer in the late 1970s and early '80s speak of the new technology's interpretive flexibility at the time of its emergence. The microcomputer marketed to consumers was an exceptionally flexible product, whose programmability permitted an enormous variety of uses. It was also flexible in the perceptions of its users. Different social groups—for instance, children and adults—might see the home computer in quite different ways. Discourses of advertising circulated a multiplicity of competing meanings. The representations in early home computer advertisements pitched the product as a Swiss Army Knife technology, with a panoply of tools for diverse purposes. Was the computer an instrument of work or mainly another way to play *Space Invaders*? Was it useful or merely amusing? Was it for adults or children? Did it belong in the kitchen, home office, living room? Was it for play, creativity, schoolwork, household management, or most basically for learning to use computers? The answer to these questions was generally: yes, the home computer was to be all of these things and more.

This was a selling point, even to the extent that many with an interest in computers, including electronics retailers, recognized the absurdity of the variety of uses promised for the new products. A cartoon published in 1977 in the retailing trade paper *AudioVideo International* captured quite well the way microchip devices were being sold as fantastical everything machines. A smiling salesman shows off an image of a computerized product with a grid of keys and a one-line readout, pointing out all of its features. Some of these are pedestrian: "time," "alarm," and "appointments." But others are more satiric: "writes novels," "babysits," "shampoos hair," "walk the dog," and "mops the kitchen floor."[58] Before computers were commercially successful as consumer goods, their pie-in-the-sky marketing was already being mocked as ingratiating and unrealistic. In the representations of advertisements, the futuristic, utopian connotations of using advanced technology productively mixed freely with the excitement of electronic play. Consumers were not asked to choose between serious and fun, but were offered the opportunity to have it both ways. A TRS-80 ad from Radio Shack pictured the sci-fi novelist and science writer Isaac Asimov seated at the computer, and the large print boasts of the product's low price, color display, and 16K of memory. The smaller print sells the computer as a futuristic, space-age, all-encompassing technology and quotes Asimov as saying, "It's like having the Cosmos at your fingertips!" The image on the screen is of Radio Shack's *Space Invaders* knockoff *Space Assault*, and Asimov is also quoted promoting its "out-of-this-world fun." But the buyer is reassured that the TRS-80 is "also a very hard-working computer" capable of managing finances, word processing, and programming in BASIC. The seriousness of a home computer could always prop up the consumer's justification for spending hundreds of dollars on an electronic toy that lets you fly a spaceship on a TV screen and shoot at alien sprites—as well as potentially untold other more important tasks. Even if its owner seldom if ever programmed a computer, organized data with it, or attempted to have it control the home's lighting and heating, the technology's potential to be used in such ways made it more legitimate than a mere game console.

Distinction was central to this rhetoric: computers were distinct from games even as they were used mainly for playing them. One important way

Figure 4.6
Isaac Asimov in an advertisement for Radio Shack's TRS-80.

that computers established an identity was through ideas about their proper placement within the home. Home computers meant the migration of electronic games from the leisure space of a media room to the labor space of a home office or a table in the home transformed into a workstation. Although both early home computers and video game consoles used television sets as displays, the computer keyboard necessitated its placement on a table or desktop, with the user seated in a chair rather than on a couch or a carpeted floor. Advertising imagery typically represented adult male users seated before the keyboard, as in the Apple II ad from 1977 set in a suburban home, with the man at the kitchen table in the foreground engaged in purposeful work with an application outputting a colorful televised line graph, and the woman in the background standing by the counter where she has been preparing food, looking over her shoulder at him contentedly. As the computer was a turnkey product rather than a kit, Apple was aiming at a new market beyond the electronics hobbyists, and showing the computer in the kitchen rather than in a workshop, in the presence of an approving and admiring woman, domesticated and familiarized the technology. But the division of labor in the home between an affluent wife and husband clearly marked the computer as a productive tool for the professional-managerial class, and its masculine breadwinners in particular. The copy on the facing page mentions games, but also assures the consumer that "Apple II is more than an advanced, infinitely flexible game machine." It emphasizes its uses for education, finance, data management, and programming, and its future potential for computer networking and composing music.

Articles about home computers in magazines like *Changing Times* echoed the image of the Radio Shack and Apple ads, picturing a man at the terminal positioned within a work environment. The next image in particular, from a 1977 story, shows the computer surrounded by paperwork, and the middle-aged white man with a mustache seated at the desk has his shirt-sleeves rolled up. The uses described in the accompanying story are mainly productive and utilitarian knowledge work: preparing a tax return, balancing a checkbook, calculating finance charges on a car loan, and helping a child with homework. Only after describing these tasks does the article mention playing chess. Later on the author admits that most of the software available on cassette at the time consists of games, implying that this might well change for the better soon enough.[59]

As these examples reveal, the idealized uses of the home computer linked this new technology to productive information-age labor, masculinizing and aging up the technology, producing associations between computing

Introducing Apple II.

Figure 4.7
Apple II advertisement in *Byte*, 1977, establishing normative gender roles for home computing.

and white-collar adult male identity. These representations acknowledged the consumer's interest in games while also reassuring them that the machine was much more than a fancy gaming device. Parents might buy a video game console mainly for children's use, but the sales pitch for home computers was more adult-centered despite the regular presence of video games and children in advertising. Just as the device was multipurpose, its market of potential users was diverse, particularly in age. Underlying the something-for-everyone appeal, however, was a consistent rhetoric of legitimacy insisting on the superiority of computers as tools to computers as

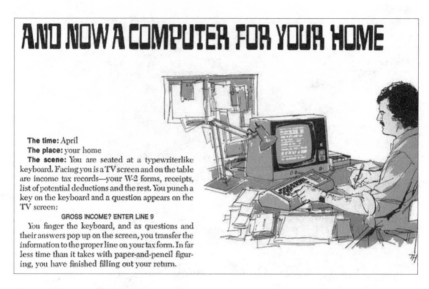

Figure 4.8
Picturing the home computer and its adult male user at work, *Changing Times*, 1977.

toys. This was particularly present in discussions of games that emphasized programming rather than playing them, as in an advertisement for the Computerland retailer in 1977 that mentioned games only as something for the computer user to program.

Commodore's VIC-20, first sold in the United States in 1980, was packaged and promoted in ways that reveal this unbalanced rhetoric of programming and other more adult uses justifying play. One of its print ads asked the consumer, "Why Buy Just a Video Game?" This was also the tagline of its television campaign with the television and movie star William Shatner, familiar for playing Captain Kirk on *Star Trek*, which emphasized the product's "computer keyboard" and the potential for consumers to "learn computing at home." At the same time, most of the images on the screen are of games. The box of a VIC-20 pictured two scenarios of use, one of a man looking over the shoulders of two boys playing a *Space Invaders* game, and the other of the same man alone using the computer for work with a cup of coffee by his side. It is notable that the adult, but not the children, appears in both scenarios. The copy on the box stresses the "arcade quality excitement" of the VIC-20 while also rattling off a list of other potential uses: household budgeting, personal improvement, student education, and financial planning.

Figure 4.9
Commodore VIC-20 advertising showing the appeal of the technology for play, with *Star Trek*'s William Shatner as pitchman.

Ads for Commodore's C64 computer, released two years later, gave greater emphasis to the diversity of uses and users. One Commodore print ad presents a grid of four photos of the same C64. At 6:00 a.m., the father in pajamas with a cup of coffee uses *Magic Desk I*, which simulates an office space representing a work station with a typewriter and a filing cabinet. At 7:00 a.m., a boy uses the computer for *Music Machine*, which is used to play music and to display the notation graphically. At 3:00 p.m., another boy plays *International Soccer*, the only video game in the ad, a baseball mitt by his side. And at 4:00 p.m., the mother sits at the computer using *The Manager*, a database application, presumably for household purposes. The copy reads: "We promise you won't use the Commodore 64 more than 24 hours a day." Here the usage of "use" rather than "play" and the presence of mainly productive or creative rather than merely amusing engagement with the computer signal the importance of the ideal of the computer as instrument for establishing its identity. The reference to hours in the day indicates the endless productive possibilities and outputs of the C64, multiplying potential functions and pleasures and linking its use to labor efficiency and management, ever defined by the logic of the clock and the segmentation of time.

Another Commodore 64 print ad shows even more various potential uses and functions, under the header "Look What's on Television Tonight,"

Figure 4.10
Commodore 64 advertising with the nuclear family sharing the home computer at different times of day.

echoing Atari's "Don't Watch TV Tonight. Play It!" Rather than using the television set for participatory games, however, the television is now made even more useful and its potential is shown to be even more diverse and multifarious. Half-hour segments of Commodore software are represented in a time grid echoing television listings in the newspaper or *TV Guide*. Some of the "programs" are games such as *The Hulk*, *Solar Fox*, *International Soccer*, and *Frenzy/Flip Flop*. Some are educational, such as *Math Facts*. Some are for work or household management, as in the earlier ad: *Magic Desk I* and *The Manager*, as well as *Easy Script*, a word processor. And there is also a Simons' *BASIC* cartridge, which one would use in programming or learning to program. The small print below touts the C64 as "the most exciting variety show on television." The idea of doing office or household work on TV so clearly expresses the value of the home computer as a new means of blurring the lines of labor and leisure, work and play, productivity and entertainment. The television becomes not merely participatory but practical and economically valuable. The computer was a new technology to integrate all of these experiences together, offering something for everyone but not in ways that necessarily evened out the values of all uses and experiences. Partly, of course, this rhetoric was a function of the need to sell an expensive product to affluent adults, eager to catch the wave of the technologically sophisticated future (more on this in chapter 5). But regardless of the commercial imperatives driving the marketing message, that message produced certain values and circulated particular meanings within a hierarchy of needs and desires. If one bought a Commodore 64 for the children, and if they were going to use it more for *Solar Fox* than *Magic Desk*, it helped to have a structure of values underlying the logic of such a major purchase. Even if they were playing games, they were also learning to use a powerful machine pregnant with social significance. Video games, "the first playthings of the information revolution," might just be the gateway to the home computer's more transformative uses.[60]

Atari Brings the Computer Age Home

The Atari 400 and 800 computers were introduced in late 1979, at the same time as Atari was establishing its place as a commercial powerhouse within the Warner Communications conglomerate, its sales revenues surging.[61] The Atari computers were marketed aggressively, though they were never as profitable as Atari's stand-alone video game console.[62] But despite

Figure 4.11
Commodore 64 advertising recalling Atari's "Don't Watch TV" campaign.

the commercial fortunes of the 400 and 800, the reputation of Atari as the leader in home video game consoles undoubtedly helped to promote the computers as technologies integrating play with other uses of computing. The technical superiority of computers for gaming made these models appealing as platforms for play while also opening up programming or other applications to home game players. In 1981, Young & Rubicam produced a series of television commercials to sell the 400 and 800 under the slogan "Atari Brings the Computer Age Home," drawing on many of the tropes familiar from marketing early home computers, but also questioning some of the distinctions between adult and child, useful and playful applications of the home computer.[63] More than other marketing campaigns, "Atari Brings the Computer Age Home" commercials exploited the interpretive flexibility of the home computer as a useful plaything, a toy with purpose. These commercials express the contradictory status of video games as both computers and gaming devices, and of computers as both adult and child technologies. They also reproduce the association, present throughout video game discourses of the 1970s and '80s, between advanced electronics and masculinity.

These television spots, like many video game advertisements of the era, place the product in the space of the suburban American family home among wealthy white consumers. But unlike typical video game ads, the meanings of these spots play on expectations about differences between not just adult and child and male and female users, but also between more and less legitimate ones. Several of them represent adults sneakily using them for play rather than work, and children demanding time on the family computer for educational use. In one ad entitled "Mom," stressing the Atari computer's function as an information hub under the mother's purview, children use an educational application to learn state capitals, while the father uses the computer for finding out scores of sports games. The dad offers his teenage son the car keys to get a chance to play Atari's *Basketball* game, and the kid turns him down. Thanks to the computer, the wife knows the score of the Bears game before her husband and has the pleasure of informing him, and the mother learns as much geography as her children, prompting a girl to say to her sister in the commercial's kicker, "I told you she's smart!" Adults use the computer to learn and play, enriching their lives and rejuvenating themselves. The voice-over concluding the spot speaks the theme of the campaign, "Atari computers: we've brought the computer age home." The appeal to the mother as much as the father or son makes for a message of inclusivity and integration of all family members' needs, though it is also telling that she is the only one in the household who does not get to use the Atari computer merely for fun.

In another commercial from this campaign, "Broker," the emphasis is on the home as a flexible space of play and work. A suburban dad descends the staircase on his way to the office, his nerdy son asking "late again?" His wife hands him a cup of coffee and Dad says, in a sing-songy voice, "I am not late." The camera tracks to follow the patriarch not out to work outside the home as we would expect, but into the study off the hallway where he hangs up his jacket and sits down at an Atari 800. The wife says, "Have a nice day at the office," and closes the door behind him. As he sits down to work, the voice-over describes his telecommuting in unambiguously positive, fantastical terms: "With a home computer and its ability to bring the world to you, some people never have to leave home to go to work." On the screen facing the man we see stock market figures, specifically the stock of Warner Communications rising in a marvelous bit of reflexive corporate wishful thinking. By selling Atari 800s, Warner and Atari were hoping that a fortune could be made in appealing to a public who might desire the integration of white-collar knowledge work into domestic routines.

However, the ad quickly shifts to other uses and users of the Atari computer. We see a hand on a joystick playing *Star Raiders*, and the reverse angle reveals the player to be a white-haired woman who pumps her fists and cheers herself and her cat on: "We got 'em, Buster!" Next we see a screen on which the computer user is calculating biorhythms, and the incongruous transition in this instance is to a tall, strong boy in athletic garb. This is a machine with such various and novel functions that it will appeal to anyone in surprising ways. Then the ad shows off stacks of Atari program cartridge boxes with diverse content: *Mortgage & Loan Analysis, Mailing List, Telelink I, Bond Analysis, Stock Analysis, Space Invaders, Touch Typing, Missile Command, Graph It, Energy Czar*, and *Star Raiders*. Such representations of the number and kinds of applications for early home computers, as with the Commodore, would seemingly avoid categorizing the programs and uses, instead representing the fluidity of more and less productive or fun uses, and to suggest that any use would carry the excitement of technological novelty and sophistication.

In the end, "Broker" returns to the father, who scans from side to side to check if anyone is looking as he prepares to play *Space Invaders* in his home office. The voice-over speaks: "The hardest part isn't *how* to use it, it's who *gets* to use it." The nerdy son reappears and Dad is caught. The boy plops his schoolbooks on the desk and asks for a turn with the computer: "Market's closed, Dad. Mind if I do some of *my* work?" The father backs away and the announcer reads the tagline, "Atari home computers: we've brought the computer age home." By reversing the interests of father and son, Atari reinforced the sense of the computer as a machine that male users in particular could not keep their hands off of, whether for serious or frivolous purposes, and insisted as well that most users would want to have it both ways.

A final example of this campaign is the spot called "Bonjour," which advertises the Atari 400 and emphasizes its educational utility. Like all of these examples, "Bonjour" includes multiple, diverse applications of the Atari computer. It begins and ends with a child learning to speak French but in the middle also represents the computer as networked to receive data and used to play *Missile Command*. In demonstrating how the computer can be used to "link up to a world of information by phone," the ad shows the computer in the kitchen connected to the receiver in the cradle of a modem. The camera pans and tilts from the screen to a man in an apron presenting a dinner platter ("Ta-dum, Chinese meatloaf!"). His wife marvels, "Recipes from the computer, great idea, George!" Just as "Broker" plays on our expectations of a young man's and older woman's interests, this one shows the man preparing the family meal, though with the technologically advanced aid of the computer. The spot also includes the obligatory shot of stacked boxes of various programs, including *Kingdom*, *Video Easel*, *3-D Tic-Tac-Toe*, *Super Breakout*, *States & Capitals*, *Space Invaders*, *Star Raiders*, *Missile Command*, *Music Composer*, *Telelink I*, and *Touch Typing*.

In the opening and closing scenes, we see intergenerational exchanges that show the Atari computer as a useful tool for young people's education and advancement. The family's son is learning French on the computer, though his father stands by to help with pronunciation. In the concluding scene, the boy's grandparents arrive at the door and, to an audience of delighted family members, the child greets them in the foyer: "Bonjour grandmère, grandpère, comment allez-vous?" ("Hello Grandma and Grandpa, how are you?") The family members surround the boy proudly as the voice-over proclaims, "We've brought the computer age home." Here the importance of home is echoed by the scenario of the family uniting, though not in play as in video game advertisements. Rather, the productive home computer brings the family together to celebrate a child's learning, a proxy for his development into a bright and successful representative of the household and also of his class and gender. American boys growing up with Atari home computers would be prepared by this technologically advanced upbringing not only to speak French, but to use computers.

As a company best known for its games, Atari was undoubtedly eager to present its computer line as more than a fancy way to play, and in creating this sense it took part is a wider discourse of early home computers as an electronic do-it-all technology. But in emphasizing work and learning, Atari also made video games into an essential part of a home computer culture that blended games with other kinds of electronic experiences, and balanced the creative, personal, pleasurable experience of computing with more instrumental ideals. The use of microprocessor devices to accomplish these quite different goals and to satisfy many distinct needs was hardly new, however, in the late 1970s and early '80s. The same combination of productive and playful functions was underlying mainframe and minicomputer game programming, techno-utopian fantasizing of a future of computers for ordinary people, and hobbyist computer and electronics culture. Early home computers made microprocessors accessible to consumers, allowing them to explore the varied uses and pleasures of this new technology. Games and play were always at the center of this.

5   Video Kids Endangered and Improved

Ten years after their emergence as a consumer product and public amusement, video games were no longer unfamiliar objects, and their novelty value was being transformed. They were increasingly present in young people's everyday lives, whether through consoles, computers, or arcade cabinets, as objects of pleasure and fascination. Although adults also played them, in the early 1980s, video games were typically seen as a form of youth culture, defining a generation in distinction to its parents. No longer were they the "new trick your TV can do" or a "space-age pinball machine," construed in a way that related the new medium to familiar technologies. Advertising no longer addressed families so much as teenage boys. The novelty now was that games had become an intense object of interest seeming to demand a bottomless supply of quarters from kids devoted, even addicted, to them. This happened only after the introduction of games like *Space Invaders* and *Pac-Man*, which rewarded obsessive repeat play and earned handsome sums for game operators.

In the early 1980s, such passion was an occasion for widespread concern among parents, teachers, community leaders and government officials, and a legion of experts in psychology, health care, and education. Some of this concern would seize on what was perceived as a serious social problem. Municipalities tried to restrict or ban video arcades. Parents and teachers were frightened by the presence in their communities of spaces believed to be corrupting youth. Video games were represented as a threat to childhood development and to the health and well-being of players. Concern about them often had strong moral overtones. However appealing they were, games were accused of being bad for kids, and bad for society. Many adults evidently subscribed to this notion wholeheartedly.

But the reaction to video games becoming ingrained in young people's lives was not only or even primarily negative. Many looked at games as a productive pursuit acquainting young people with advanced technology and preparing them for future engagement with computers, especially in the workplace. For every complaint about video games leading to crime, sex, or addiction, one might hear a very different kind of expression indicating the potential for electronic play to inculcate useful skills and habits, to be educational or therapeutic. Video games were seen as a herald of the computer age and the information society. While some feared or disdained the computer as an instrument of control and dehumanization, many others saw its inevitable transformation of work and leisure in welcome and productive, even utopian, ways, as we saw in chapter 4. They expected young people to benefit from familiarity with computers along with other kinds of electronics and telecommunications.

What these two kinds of reaction to video games, the fearful and the hopeful, had in common was a conviction that the emerging medium would have a profound and long-term impact on the young players so enamored of it. Whether harmful or productive, the games were sure to affect their young aficionados. The novelty of video games as a generation's defining pastime brought with it a common belief that games were sure to shape young people's whole lives.

Whatever effects they may have had, video games could not have been as powerful as they were given credit or blame for being. The ways they were understood in the 1980s were products more of imagined than real effects. As they became commonplace and enormously popular, video games also became sources of intense fascination and fear. Instead of seeing what kids were actually doing with and around video games—having fun, competing with peers, learning mastery, socializing—many unsettled adult observers used them as objects on which to project fantasies of endangerment or improvement. As Carolyn Marvin has argued, "new media embody the possibility that accustomed orders are in jeopardy."[1] New media seem to speak to a culture's hopes but also its anxieties, prompting ambivalence about the structure of society and its future prospects.[2] Decades later, the panic and outrage over video games, and efforts to regulate and proscribe them, may figure more into popular historical memory, but at the time this was only one facet of the response to electronic play. Another, equally prominent in public discourse, was much more optimistic. Both the fearful and hopeful reactions can be seen as expressions of the same underlying concerns about changing social roles and shifting expectations about family, leisure, and work. They were both ways of coping with the novelty of electronic play in the lives of a generation coming of age in a new world.

"Trouble in River City"

One public venue for expressions of hopes and fears about video games was the popular press, including print and broadcast news. News programs and stories often presented the impact of games as a matter of ongoing controversy and debate. Following the convention of much North American journalism, news on video games was framed as a controversial issue with two opposed and equivalent sides. Representing one side might be a municipal office holder, PTA representative, teacher or parent, speaking of the harms inflicted or liable to afflict the town. For the other might be an arcade proprietor, social scientist, or real-life video kid testifying that the games were at worst benign fun but also possibly a useful form of leisure. Unlike some moral panics stoked by fear mongering in the popular press, video games were often shown in a somewhat balanced fashion, with proponents and opponents lining up against one another. The unfolding of court battles, such as the case of Mesquite, Texas, whose regulation of coin-operated games went all the way to the Supreme Court in 1982, meant that reporting could easily follow a he-said/she-said format.[3] This might still have given substantial credence to the con side, whose fears were quite a bit overblown and often wanting for evidence and logic. But they also countered it with a pro point of view, or at least a position skeptical of the harms attributed to the medium.

As an example of the journalistic framing of video games as a hot topic with paired sides, consider one trope applied on several occasions in discussions of video game effects: the notion that video games were prompting new "trouble in River City."[4] This was an invocation of *The Music Man*, the 1957 Broadway musical adapted to film in 1962. Now, the news might report, the trouble isn't "with a capital T that rhymes with P that stands for pool," as in the famous show tune, but with a V. Readers or audiences were thus reminded of the fictional all-American place where people were easily persuaded by the fast-talking traveling salesman Harold Hill that the presence of a new pool hall posed a threat to the town's young people. This new threat was associated also with youth styles of dress and speech (knickerbockers too low, "swell"), and was sure to lead to all manner of vice. The trouble could be removed by interesting youngsters in the band instruments and uniforms that Hill was selling, a more wholesome and communal form of cultural expression.

Perhaps the readers and audiences of news stories mentioning "trouble in River City" would not have recalled the whole scenario of *The Music Man*. But those who knew it might remember that the trouble was totally concocted, that Hill was a film-flammer out to take River City's money and then slip out of town, and that the nostalgic representation showed that citizens were too credulous in accepting claims that pool poses a danger. *The Music Man* does not stoke the audience's fears of children's newest amusements. Actually, it affectionately mocks popular panics by showing bygone small-town mores, congratulating the audiences of the 1950s and '60s for their greater sophistication and modernity. So by referencing *The Music Man*, news items indicated not so much that video games were as harmful as hard drugs or as likely to lead to criminal activity, as that they were an object of concern among cautious and worried ordinary good folks who might too easily be convinced of their danger. That is how things happened in River City.

Harold Hill's crusade in *The Music Man* revives a familiar scenario of youth in danger: "keep the young ones moral after school," the song goes. Parents naturally feel concern for their children's well-being, and often too readily accept that some novelty that catches their interest must be a hazard. The history of moral panics around young people's leisure-time interests and behaviors shows a remarkable continuity across generations. This follows a pattern of historical amnesia, as one generation forgets the same expressions of fears articulated by its parents not so long before about the novelties of its own youth.[5] Sometimes these panics are not media specific. But their representation in the media not only brings them to consciousness, it also stokes their fire, helping them to spread and grow. Sometimes the objects of these panics are themselves new media technologies, like motion pictures or the Internet. Sometimes they are media genres, like dime novels or comic books. Media panic, a subtype of moral panic in which technologies, formats, or genres of media are at the center of public outrage, describes the widespread feelings of fear and anxiety around video games in the 1980s.[6]

Moral and media panics alike identify a form of youth culture seen as deviant and dangerous, whose threat is blown out of proportion if not invented entirely by the guardians of virtue. The object in question functions as the scapegoat or folk devil to which responsibility can be assigned.[7] A panic is a moment in which generational tension bursts open. It exposes the pain of parents losing the world of their youth, and finding their own children's lives unfamiliar. This tension and trauma spurs moralists to emotional excess and polarized, panicked expressions. It pits reasonable adult authority against a heedless youth gone wild, against their own children overcome by the effects of new media. Of course the reverse is also true: panicked reactions to new media in the name of reason tend to be emotionally excessive, and children's defense of their fun can be remarkably cool-headed. In such cases, logic tends to be on the side of the young.[8]

Many observers of media panic see the specific object placed under intense scrutiny to be a substitution for matters of deeper societal concern. Kristen Drotner argues: "Panics are deeply implicated in political issues beyond their immediate causes."[9] Dmitri Williams regards the panic over video games to be an expression of concern over changing family dynamics as women entered the workforce and children's unsupervised leisure time seemed to be a problem in relation to emerging sexual roles in the family.[10] We can additionally see the upscaling of coin-operated amusements in plush family fun centers and suburban game rooms to carry along unwanted old working-class connotations, and arcades might be seen as schools of vice, unsuitable for middle-class offspring (see chapter 1). Frequently, as in *The Music Man*, a low-culture form of amusement is placed in opposition to high culture or at least middle-class culture, and the low form is presented as an enemy of the people, in effect as a proxy for fears of failure to reproduce social class. Complaints that games were addictive or crime-related function as well as a cover for economic and cultural anxieties.

Adults also fastened onto the video game as an example of rampantly advancing technology, a form of computing with which children but not their parents would become intimately familiar. They would have experienced this as an incursion of the technology of the new information society, so commonly associated with advanced electronics, into their children's leisure. The world their sons and daughters were inheriting was in some ways markedly different from the postwar baby boomers' environment. It was of course uncertain whether exposure to video games would teach young people how to work effectively with technology, or whether the technology so touted as beneficial might instead turn out to be harmful to their mental health, sucking them into microworlds and isolating them from society, influencing them to become bellicose and short-tempered, transforming them into zombies unable to function outside of the highly mediated psychic environment of the game. There was no shortage of scary scenarios and potential threats.

Addiction, Violence, and Media

The media panic around video games in the early 1980s sometimes spread word of vague harms. There was a general feeling of contempt expressed in the usual kind of condemnations of new media, as in one letter to the editor of the *New York Times* asserting that video games were "cultivating a generation of mindless, ill-tempered adolescents."[11]

But such thinking could also be quite detailed in portraying the trouble at hand. Among the specific dangers attributed to video games in publications of the time were both minor and major problems, including physical ills such as "Space Invaders Wrist," skin, muscle, tendon, or joint problems in the hands, and damage to the eyes.[12] Squandered lunch money was a frequent point of concern, which suggested that some children were starving themselves for *Centipede*. Worse yet: to have your lunch money stolen by a *Frogger* fiend.[13] Good kids might be so seduced by electronic games that they would resort to theft to feed their habit.

Adults sometimes feared the clusters of teenage video game players (mainly boys) hanging around stores and shopping centers, who might be perceived as obnoxious or menacing, and who might sometimes harass passersby. A CBS News broadcast on July 9, 1982, quoted a municipal official in Boston describing youth congregating in a laundromat by the video game machine, offering "fast remarks" as adults passed by and terrorizing customers. In popular media discussions of the day we find expressions of fear about video games being linked not only to this irritating or hostile behavior like loitering or harassment, but also to truancy, panhandling, vandalism, gambling, loansharking, and taking or selling drugs. In some instances, adults feared possible pathways from video games to illicit sex. Children might be committing crimes to finance their habit, and some adults wondered where all those quarters came from ("Just how much help do the neighbors need each week?").[14] As one newspaper reporter summed up the generational divide over games, "To many parents, the glowing, beeping machines are Molochs to whom their children are lost. They conjure up fears of purse-snatching, truancy, and '42d Street drug dealing.'"[15]

Young patrons of the arcade were standardly described as mesmerized addicts, and sometimes their endless appetite for video games was likened to other addictions: caffeine, cigarettes, alcohol, heroin. Video games were routinely called addictive as if that were already a matter of established fact.[16] Children also adopted this rhetoric in quotes to the press by calling themselves video game junkies or addicts, though their usage could be playful and ironic rather than alarmed. This addiction would explain the truancy and petty theft supposedly caused by games: only addicts would go so far. An item from Mesquite, Texas, an epicenter of video game panic, reported that addicted kids would break into cigarette vending machines to feed their cravings, pocketing the quarters and leaving behind the nickels and dimes.[17]

Television, another form of popular media blamed for negative effects on children, and likewise seen as analogous to drugs, was a reliable comparison. Video games replaced TV in some players' leisure time, and were played in the home using a TV set. Some observers wondered if video games were even more addictive than TV owing to their interactive, participatory nature. A letter from a psychotherapist published *Psychology Today* in 1983 was given the matter-of-fact headline "Addictive Video Games."[18] Just like TV, it was feared, video games could turn out to be an impediment to children's success in school and life.

Even if the children were not becoming victims or perpetrators of crimes thanks to video games, many adults worried about the new medium's effects on young players' cognitive and emotional development. In informal comments following a speech on the subject of violence in the family he delivered in Pittsburgh on November 9, 1982, the US Surgeon General, C. Everett Koop, asserted that children were becoming addicted to video games "body and soul," and suffering adverse physical and mental effects. His comments made international news and stirred controversy. He walked them back the following day, clarifying that video games are no threat to children, but fears had by then been aroused.[19]

The attention commanded by this controversy and reversal suggests that the violent content of popular games was occupying many adults' thoughts. Just as television violence was believed to encourage young people to act out real-world violence, video game violence was similarly a cause of worry about aggression and acting out. So many of the popular games at the time involved shooting and killing in heroic scenarios of invasion or war. The difference again between TV's perceived passivity and video games' interactivity was a further impetus to fear the impact of the new medium as a stronger and more damaging force than the more familiar small screen. Like TV, video games could also be seen as an isolating influence, a substitute for socializing. As an even more engrossing technology, games might be even more threatening to childhood sociability than television. Their association with a medium already often viewed with alarm and outrage in the late 1970s and early '80s framed video games as a social problem.[20]

Moral Entrepreneurs and the Regulation of Play

The video arcades sprouting up all around the United States and many other countries in the 1970s and '80s became magnets for negative attention in numerous communities. The campaign to do something about the video games found in these arcades, and their putative harms, was waged in local municipalities as well as in national or international media. Some of the citizens of a city or town might view the novel presence or the proposed opening of a video arcade as an invitation to alarm. Video games would likely have already been present not only in homes but in the many public places that added arcade cabinets in the 1970s and '80s such as convenience stores, supermarkets, bowling alleys, pizza parlors, and bus depots, but arcades were viewed with particular suspicion as hazardous spaces for children. Citizens who saw arcades as a threat to the youth of their locale might have summoned any available evidence and rhetoric to mobilize for their campaign.

Their crusade led to the renewed regulation of public amusements, such as ordinances barring children under eighteen from playing during school hours or requiring their accompaniment by a parent or guardian. In some municipalities citizens tried to prevent new arcades from opening at all by applying zoning by-laws, limiting the number of machines allowed in a single establishment, or passing ordinances to regulate electronic or coin-operated game machines. Some of the municipal ordinances covering electronic games already existed for coin-operated amusements such as pinball. These various efforts quite often led to local conflict over video game regulation, including a number of legal cases between municipalities and business owners.[21] Often appearing in items in the newspaper or on television, these cases made video games newsworthy, providing a hook for a story about the "two sides" of the newly popular medium.[22]

Efforts to ban video games, or at least to protect children from them, looked for authority to educators, law enforcement officials, business leaders, mental health professionals, and other concerned members of the community. Anyone with a degree of social prominence or power might give weight to public statements opposing the games. These moral entrepreneurs often gave interviews to the media or presented themselves publicly as voices troubled by video games and their potential to harm their players.[23] The media panic over games drew for its energy and credibility on these public persons defending the social order and status quo against a threatening, corrupting newcomer.[24] They expressed an essentially conservative social ideal in which the community would remain unchanged by developments in technology and popular culture, and would protect its established ways against the seductive appeals of dangerous novelty. Moral entrepreneurs are prominent citizens whose concern over what they perceive to be deviance from community values leads officials of the state to make or adapt rules to proscribe or forbid objectionable behaviors. No mere ordinary members of the community, they claim superior knowledge and powers of judgment. They recognize a hazard, want it to be managed or stopped, and speak out forcefully. Video arcade moral entrepreneurs in American towns were often associated with schools or government. Some became outraged beyond all reason. Despite the authority they claimed, these antigame activists typically offered no compelling evidence in support of their position, but rather hysterical fantasies of youth gone wrong and overwhelming technology.

By 1982, the economic success of video games led to the proliferation of arcade cabinets in public places. Installing new arcade games was often seen as a legitimate get-rich-quick scheme, or at least as a smart business venture. The regulatory battles against video games were often waged in middle-class suburbs like Centereach, New York (on Long Island), Morton Grove, Illinois (north of Chicago), and Mesquite, Texas (in the Dallas-Fort Worth metroplex). On one side were entrepreneurs or big companies like Bally, which ran the Aladdin's Castle arcade chain, along with the teenagers of the town who felt persecuted by crusaders against their fun. On the other side were the parents, moral entrepreneurs, and municipal officials. Trade in video games was unlikely to be regulated at the federal or state level, and by the early 1980s game consoles were finding their way into millions of homes alongside the television set. But municipal governments could adopt regulations to control the placement and use of video games in public, and it was around these proposed or adopted ordinances that the most public frictions occurred. As a visible battle pitting two camps, each with its own form of power (often political power versus economic or cultural power), the struggle to regulate or not regulate video games in these American municipalities made for a conventionally framed news item.

Some common forms of ordinance adopted in the early 1980s included restrictions on the number of coin-operated electronic game machines one establishment could have, restrictions on the age of patrons in game rooms (and perhaps variations in this policy by time of day), restrictions on children entering game rooms unaccompanied by a parent or guardian, and restrictions on game rooms opening within a certain proximity of a school. By adopting zoning regulations, a municipality could keep arcades away from certain parts of town. Unless the games were kept out of the city altogether, as happened in Marlborough, Massachusetts, and Coral Gables, Florida, these policies regulated environments of play and ages of patrons. They could not, however, prevent children and teens from playing video games altogether. Local ordinances were an effort to curb and curtail the attraction of games and their presence in young people's lives. They had other perhaps more significant functions, in particular regulating the behavior of teenagers in public places. Still, passing ordinances to regulate game rooms was often portrayed as necessary to counteract the harmful effects of the medium on children. Naturally, these effects were merely asserted rather than shown to have the support of empirical research, and in one legal case, an expert for the defense testified that there were no such demonstrated effects.[25] Whether there actually were harmful or helpful effects, each side was invested in a set of ideas about video games that served their interests.

Moral entrepreneurs agitating against games accused them of getting kids hooked, and if not causing full-on physical addiction, then at least hypnotizing young people in a way that would make them uninterested in anything else.[26] Often this focus on the strength of electronic games' appeal was a way of expressing fears that children's fascination with games would spell failure in school. Education under threat was a frequent theme in the fearful discourses of the antigame crusaders. Kids were believed to be ditching school to go play video games. They dropped their lunch money, meant for the school cafeteria, into the slots. Fears of truancy caused by arcades led to ordinances barring children from game rooms during school hours, and requiring that any arcade be 1,000, 2,000, or even 2,500 feet from a school. But even outside of the school day, kids' inability to stop playing was seen as an impediment to proper development; a fifteen-year-old child out playing video games at 10:30 on a school night might be unlikely to do well the next morning, causing concern for his parents.[27] Without any evidence that players of video games did any differently in school than their peers, the members of city councils and other local governments restricted their use in the name of protecting students from failure.

While the games themselves could be considered an addiction, they were also seen as a gateway to substance abuse. The mayor of Bradley, Illinois, a town that made national news for its video arcade regulations, claimed that hundreds of kids had been seen smoking marijuana at a game room nearby. Public arcades were, in many parents' fears, the place where children on the straight-and-narrow path would be given drugs.[28] The National PTA, along with many concerned citizens and local governments, worried in particular about public game spaces in which supervision and regulation might be much too lax.[29] As hubs of teenage socializing, such unruly locations were seen as likely gang hangouts, trading posts for drug dealers, and "dens" in which children would find access to booze and dope. The successful movement to ban video games in Marshfield, Massachusetts, was spearheaded by a retired police officer who feared games in public places would lead to crime and drug use, a concern in many towns.[30] Some of the strongest fears expressed about video games seized on the dangers of the public space as much or more than the dangers of the medium. But typically the panicked expressions of moral entrepreneurs picked up on both of these sides of the fearful reaction against electronic play.

Those afraid of games themselves would tend to see their violent themes as troublesome. Just as television was thought to inculcate aggression or violence in children, so video games were seen as a likely cause of antisocial behavior and desensitization. While sometimes the interactivity of games was seen as a positive attribute by comparison to TV's passivity, when thinking of violence, a critic could say that interactivity makes games' effects potentially more detrimental than television's. The player is the one committing the acts of shooting and killing, turning leisure-time diversion into practice for war. Many parents and local leaders feared that video games were breeding a violent generation likely to endure "long-term psychological damage" from *Space Invaders* and its ilk.[31] In one article on the controversy over effects of games in the *New York Times*, a professor of psychiatry, psychology, and pediatrics at Rutgers University, Dr. Michael Lewis, worried about the intensity of the experience and the sense that the violent scenarios in games are real rather than imaginary.[32] A 1983 article in *Health* by a pair of Bethesda, Maryland, psychotherapists, expressed the conviction that like troublesome representations on television, violence in games can lead to violence in real life.[33]

The Bethesda therapists also expressed concern over a tendency observed by adults that games "isolate" children and harm their socialization and development. CBS News reported on January 29, 1982, that "psychologists are beginning to worry that some youths are becoming spaced out on the space games," leading to introversion among serious players. Like violence and truancy, isolation could pose a danger to the child's academic achievement and maturation into a responsible adult. Computers more generally were often seen in these years as being so absorbing as to take the user away from human interaction into a strictly "man–machine" encounter, a quality, as we have seen, that the MIT professor Sherry Turkle called "holding power."[34] This was its own cause for worry. But in enveloping youngsters in violent fantasy worlds, video games in particular were feared as a horrible escape that would ruin the minds of the young and halt their progress toward responsible adulthood.

In some expressions of outrage or concern, many of these themes would be woven together. Television and addiction were already familiar friends, as in the title of Marie Winn's popular screed *The Plug-In Drug*, as were television and violence.[35] Video games fit into a familiar script of concern among adults around children's leisure-time pursuits. A school principal's statement in a 1983 ABC *Nightline* episode on video games sums up this interweaving of fearful sentiment by moral entrepreneurs:

> The problem is that we have television, we have television violence. We now have a video craze where the theme of these games is violence. I think it's another ripple effect of our standards. We've had certain standards and they are slowly eroding. And there are those of us who believe that this is just a symptom, the video games craze is a symptom of another addiction that's taking place in this country. I use my crystal ball as a father and as a principal of twenty-seven years in the business of education, that if we don't do something about this it will be another nail in the coffin of our country.[36]

As in many of the outraged expressions of moral entrepreneurs, it is not entirely clear what effects the principal is attributing to the technology or its specific forms or uses. The fear of decline, of being on a course to disaster, seems to come from a source much deeper than video games.

This sky-is-falling rhetoric can be hard to square with the limited powers exercised by municipal governments in the legal or legislative matters that made news. Since games could not be outlawed from millions of homes, local governments found themselves regulating technology that was going to be part of some citizens' everyday lives no matter what. Most of the regulatory efforts were aimed at keeping children away from games during school hours, and some were aimed at keeping large game rooms from opening up that might serve as teenage hangouts, while leaving video games in restaurants or stores as they were. Keeping arcades out of town might have a different function from keeping kids away from video games during the day. The arcades were seen as trouble spots for reasons beyond the appeal of the games within them.

In 1983, Morton Grove, a Chicago suburb, opposed the opening of an Aladdin's Castle emporium with one hundred arcade machines at a main intersection of the village. An assistant village administrator quoted in a *Tribune* story about the conflict over the arcade conceded that games might not be bringing about "the downfall of civilization." But he insisted that the business would cause "headaches."[37] The village board voted to deny Bally from proceeding with its plan in February 1983, keeping in place an ordinance forbidding businesses from operating more than ten game machines. The *Tribune* coverage of this local government action noted that the board "foresaw large groups of teenagers loitering in front of neighborhood stores, possibly drinking or causing disturbances, if game arcades were permitted in the village."[38] In towns such as Westport, Connecticut, zoning and planning officials worried that arcades would attract crowds, create parking problems, and demand the attention of police.[39] In well-to-do Cliffside Park, New Jersey, as in many genteel suburban municipalities, city councilors feared youth from out of town congregating at the arcade, "lurking, littering, and leering at shoppers."[40] Local arcade ordinances and other municipal policies and actions were much more geared toward regulating children and teenagers than they were regulating games as such. They aimed to protect the peaceful and calm atmosphere of small towns and suburbs from the potential disturbance of young people at leisure who were perceived as unpleasant or dangerous.

A form of class anxiety at the root of antigame efforts is also evident in the example of successful arcades that adapted to the suburban realm of parental concern, like the plush family fun centers and shopping mall arcades described in chapter 1. Westport was one town where the local Planning and Zoning Commission sought to prevent an arcade from opening. This was a "luxurious video game palace," as described by the *New York Times*, called Arnie's Place. To assuage local fears of youth being corrupted, Arnie's Place was staffed with attendants in blazers and outfitted with a dozen CCTV cameras and monitors to keep a constant eye on the premises. A public address system was used to announce names of children called home for dinner, and late in the afternoon the time of day was regularly announced to remind young patrons of the hour. Parents of the town trusted that their children were in a wholesome environment free from ruffians and hoodlums, a place where they were unlikely to be exposed to mind-altering substances, sex, or crime. Kids playing at Arnie's after school would responsibly return home on schedule rather than stay out till all hours doing God knows what. The values of bourgeois suburbia and the business interests of the entrepreneur who opened the game room were made to fit together in Westport.[41]

Making arcades into safe spaces was one way for entrepreneurs and the coin-operated amusements trade more generally to negotiate concerns about video games and their effects without losing business, but another was to sue the city or town. When video games cases appeared in court, judges often looked favorably on the concerns of municipalities to protect their order and calm. Several cases concerning zoning regulations applied to video games made their way through the courts, which often found that restricting access to minors and keeping large game rooms from opening was not a violation of the First Amendment protections of speech or assembly.[42] In a 1982 New York case, *America's Best Family Showplace v. City of New York*, the court ruled that the city could regulate games in order to curb noise and congestion, finding that video games are not a form of speech protected under the First Amendment.[43] A 1983 Massachusetts decision in *Malden Amusement Company v. City of Malden* included the defense of a "legitimate objective of maintaining order, preventing crowding, and diminishing the prospects of out-of-town people congregating" in the town.[44] The America's Best Family Showplace decision was cited in the Malden decision as a precedent for denying video games First Amendment protections.

As has been true of other emerging forms of low or popular culture such as cinema and comic books, the Supreme Court was slow to recognize games as a form of protected speech.[45] It declined to affirm this protection in the Mesquite, Texas case about restricting the age of unaccompanied patrons at Aladdin's Castle. The Supreme Court never considered the case's First Amendment implications; it sent the matter back to the lower court (which had found video games to be protected speech).[46] Marshfield, Massachusetts, was one rare locale in the United States to succeed at banning video games altogether from public places, and while Marshfield's regulation did make its way through the courts, the Supreme Court declined to hear the case, effectively letting the ban stand through inaction.[47] Without weighing in on the harms or benefits of playing video games, the federal courts still recognized the rights of local governments to regulate new media and technology in keeping with their rather conservative values, and sometimes conveyed skepticism about the potential of games to be expressive or to convey ideas and information. Perhaps similar regulations would have been accepted of, for instance, libraries or bookstores, but the threat associated with video games and youth culture made their public venues much more likely to be magnets for dispute and regulation. Had the concern over games been more widely recognized as an instance of outraged panic rather than part of a balanced debate about the new medium's effects, perhaps efforts at local regulation would not have been so successful. Yet, however much success they won, moral entrepreneurs and local governments serving their interests ultimately did little to prevent the generation coming of age with *Space Invaders*, *Pac-Man*, and *Donkey Kong* from playing video games and identifying so strongly as their players, as video kids. The moral entrepreneurs and the press who lavished them with attention did succeed at making games into an object of controversy, though, producing a perception of danger and disrepute in tension with another, more productive and positive reputation.

Pac-Man or Perish

In framing the new medium as an issue of two embattled sides, stories about video games often looked to experts capable of speaking to their harmless or even salutary qualities. Many news and magazine items also aimed to defuse media panic and paranoia, or to explore the potential of electronic games to be a gateway to more advanced and sophisticated computing.[48] Some young people, for instance, would graduate to programming rather than merely playing games. Playing games on computers would be a first step toward learning the inner workings of the new machines, as we saw in chapter 4. These stories appeared as newspaper op-eds, features or columns in intellectual publications like *Smithsonian*, or popularizing ones like *Psychology Today*, and in coverage of scholarly research. Several books by academic experts were also published in the early 1980s intended for general readers expressing hopeful and inspiring ideas about computers and games as technological advances. Psychologists, sociologists, education researchers, healthcare professionals, and others with clinical, scientific, or academic credentials spoke and wrote publicly both to diagnose the excessive fears of the moral entrepreneurs as a hysterical overreaction, and to identify benefits accruing to young people playing electronic games.[49]

To many intellectuals appearing in popular media—in contrast to typical concerned parents, teachers, and local officials—video games fit into a larger narrative of technological development, typically conceived as progress toward a new society taking shape. Rather than regard video games as the latest in a history of problematic public amusements ("Trouble in River City"), these professional thinkers looked at them as the herald of a transformed society. A new day would mean new ways of thinking and being, and machines would show these new ways to us, functioning as aids to human imagination and work in the new society. As children, the main market segment for video games was poised to learn from them habits, skills, and an understanding of technology that some adults felt sure would lead to a successful future of schooling and work with advanced electronics. Video games might have appeared to some adults like silly, trivial diversions about zapping aliens with a ray gun, but to optimistic experts they were "the first playthings of the information revolution."[50]

The idea that video games were one piece of a much larger economic and sociocultural transformation was rooted in at least a decade of public discussion about the development and future of American life. Writers including the sociologist Daniel Bell and the political scientist Zbigniew Brzezinski had written and spoken from their prominent perches as public intellectuals about a shift underway from modern industrial society to its successor, which Bell famously termed "post-industrial society."[51] Industrial society, based on the Fordist production of goods, was by this account giving way to a new service economy in which information would be more central than material production. In his grand scheme of history, Bell saw three periods: agrarian society, based on farming; industrial society, based on the nineteenth- and twentieth-century technologies of factory manufacturing; and postindustrial society, based on knowledge work made possible in part by electronics. Computers were not the only new technologies of the twentieth century to figure into such thinking. For Brzezinski, electronics more generally was the notable innovation, and he called the new age "technetronic." Cable television (including two-way cable TV), satellite systems, digital telephony, and telecommunications more generally were also central in much of the thinking about large-scale economic shifts during the 1970s. In one utopian futurecasting account, the prefix "tele" did much of the rhetorical work of setting the stage of a new society in which education becomes teleducation, medicine becomes telemedicine, and so on.[52] "Cyber" would eventually carry similar rhetorical weight, but "tele" was more familiar at the time.

By the 1980s, computers, another electronic technology, were regarded as most central to the changes underway. Telecommunications could collapse distance and produce a more participatory mode of engagement among users, but computers could enable knowledge work of many kinds and apply systems theory to solving complex problems in large institutions of business and government. The rise and spread of this type of knowledge work, according to Bell, would transform the countries of the West from blue-collar to white-collar societies, with a professional and technical class soon becoming the largest group in the labor force.[53] In this new age, "What counts is not raw muscle power, or energy, but information. The central person is the professional, for he is equipped, by his education and training, to provide the kinds of skills which are increasingly demanded in the post-industrial society."[54]

The Coming of Post-Industrial Society, Bell's 1973 book, was the origin point of many ideas about this transformation, and was widely reviewed and discussed. Bell had spoken publicly on its theme first in 1959 and throughout the 1960s, so the ideas were very familiar in learned circles. In many publications in the 1970s and early '80s, from news items to popular nonfiction books, the coming of the postindustrial society was taken as a given, a fact of life, and the shift toward knowledge work and an information economy was a commonly shared vision of the present as it was becoming the future.[55] In this future, the computer—the thinking machine—was making possible a world in which knowledge and technology are the most precious resources. To bring up a workforce adept at using computers, education would need to be expanded, with college becoming more available to more young people, and more necessary for social mobility. Professional labor would demand educated citizens trained to work in fields such as mathematics, engineering, finance, health care, and many others requiring technical skills. The society's elite would comprise a range of highly educated experts in lines of knowledge-based service work. As one newspaper trend story on the service economy put it, "the future ... is in offices, not factories."[56]

To the optimists pushing this vision of the new age, the computer would be a strong force for significant change.[57] To prepare for and take part in the emerging society, many people were learning computer skills even if they had no clear immediate use for them. "Computer literacy" was a term gaining currency, with its connotation that knowing how to use a computer is comparable to being able to read and write.[58] A 1983 newspaper story summarizing many of the key points of John Naisbitt's bestseller *Megatrends*, one of the books positing that knowledge work is replacing industry, expressed a commonplace assumption that more teachers of computer skills would be needed in the future.[59] Social scientists at the time believed that the microcomputer would be to the information society what the automobile had been to the industrial society, and that video games were teaching children that microcomputers were easy and fun, acquainting them with essential technology.[60] A 1983 report from a conference on children and computers discussed their utility as educational technologies in their own right. Headlines used phrases like "computer age" and "information age" interchangeably and routinely, assuming the reader would find them to be familiar shorthand and aptly descriptive of the changing times. For children growing up in a "computerized society," wrote one psychologist in a letter to the editor of the *New York Times*—reacting to the media panic around arcades in Centereach, Long Island—video games could turn out to be "the educational tools of tomorrow."[61]

With rhetoric like this, the experts defending video games championed their potential to breed familiarity and facility with essential technology.

Children by these years were frequently encouraged or assigned to use computers in school or at home.[62] Kids attended computer camps and took computer classes, and for middle-class children in particular, using computers was seen as an enriching experience that could develop a useful skill. American schools had 150,000 computers in 1983.[63] An ABC News story airing July 8, 1982, reported from a summertime computer camp on a college campus where children were learning from experts and also exploring the technology on their own. A mathematics PhD, one of the camp's counselors, said of the campers: "They'll be able to teach me fifteen years from now ... they'll be ready for the computer age." Many researchers were studying the educational applications of computers, including the potential for games to be integrated into educational software. Computers used for play, what Sherry Turkle called "high-tech rec," could have the double function of being both a diversion and a pedagogical tool.[64] And researchers were eager to extract lessons from video games to understand what makes them so much fun, the better to design effective educational software.[65]

Even without considering titles written for educational use, the video games popular with young people in the early 1980s were expected to yield long-term benefits to devoted players. The kids spending so many hours mastering *Asteroids* or *Ms. Pac-Man* were obviously engaged in learning. A frequently offered example of video game learning in popular discourse was eye-hand coordination, which was believed to be learned through regular arcade play.[66] Whether this was to be a useful skill in the information age was rarely addressed by those making such claims; actually they were often made dismissively, as though quick hands were the only benefit of video games to line up against a long list of harms. But on the positive side, in addition to honing reflexes and sustaining attention, games demanded that players develop detailed understanding of their patterns and play structures, inculcating a feeling of mastery over a sophisticated machine. Proficient players were conquering technology.[67] Electronic games exploited the child's natural capacity to learn, to learn through play, as in the theories of John Dewey and Jean Piaget.[68] The science-fiction author and technology maven Isaac Asimov said that games are "an important (perhaps unequaled) teaching device."[69] Video games were giving children motivation to educate themselves in the use of computers, since games were fun and rewarding.

Without even intending to, video game players were giving themselves a leg up on peers less familiar with this newly essential technology, gaining an edge in the information age. Young people growing up with video games would naturally have an advantage over their elders thanks to this intimate childhood familiarity. Like learning a language in childhood, learning computing while growing up would come more easily and become more fundamental to one's habits of thought. It could even be harmful to children to avoid or fear computers, so encountering them at a crucial period of development, in the form of video games, would help lead to future success. Apprehension about games' negative effects ran up against a notion of their becoming virtually essential to maturation in the information society. As Turkle told *People* magazine in 1983, "To feel afraid of computers in the next decade is to feel afraid of life."[70]

One social scientist at this time doing ethnographic video game research found that affluent families in the San Francisco Bay Area acquired game consoles specifically as a way of accessing computers. Edna Mitchell, an education professor at Mills College in Oakland, studied family dynamics in the home around video games. In her study, supported with funding from Atari Educational Foundation, she observed that children did not become video game "junkies" (though one divorced mother did) and that family life was often enhanced by common experiences of electronic play. "Families reported playing together, interacting both competitively and cooperatively, communicating, and enjoying each other in a new style." While families' experience of games might have been mainly for fun, the purpose of the acquisition would have been justified by loftier ambitions. "The video-games, which were often advertised as computer-related, were seen as a way of familiarizing children with a new and important technology. Adults were interested in being part of that learning, and the fact that these were games, rather than real work, made it easier for the adults to participate." Mitchell also observed that as computers became more widespread in American homes, it was more likely that well-off families like those in her study would upgrade from a game console to a home computer. The more the experience of games approached computing, the more the games would be seen in terms of learning rather than amusement.[71]

Computer games were no mere alternative to conventional classroom education. In many expert accounts, learning from games or computers could improve in important ways on traditional learning. Rather than passively absorbing information, computers would provide for active engagement with concepts and processes. Video games made learning computing fun, and children needed no encouragement or discipline to engage with them. Games were compelling, inherently gratifying activities. They were intrinsically motivating to their players: mastering them would be its own reward, and it encouraged further learning. With computer games, "you will be learning without even knowing you're learning," according to Isaac Asimov, "because we don't call it learning when we are doing something we want to do."[72]

As in many fantasies of computerization, machines would improve on tasks previously performed by people, working for us by being our betters, and these computers would at the same time teach us lessons and help us learn to use them. A college professor told *Smithsonian* magazine in 1982 that "the arcade machines are skillful teachers ... with more variety and individuality than many human teachers."[73] A University of Michigan professor observed in the same story that the kids in the campus arcade seemed much more interested than those in the library. Learning with computers and also learning about computers were regarded as essential to an information-age education. The same commercially booming arcade cabinets seen by moral crusaders as the devil's playthings were often imagined by their academic champions as being at the center of an emerging new pedagogy. The *Smithsonian* author summed up this line of thinking with a provocative, not totally incredulous question: "Are these psychologists saying the arcades are the new schools for survival? Is it somehow more valuable to learn Missile Command than to learn English?"[74]

Programming games was the most obvious productive use of computers to come from playing games. Geoffrey and Elizabeth Loftus's 1983 book *Mind at Play: The Psychology of Video Games*, emphasized this desirable benefit. The Loftuses were married psychologists, and their account was one of several popular books published within a brief span of time in these years by academic researchers giving their endorsement (however qualified) of the value of the new medium.[75] Loftus and Loftus saw video games as "potentially the most powerful educational tools ever invented."[76] They emphasized that video games would be valuable in introducing young players to computers. They also noted that young people by 1983 had already made the transition from playing to programming, and anticipated that some of these young programmers would find employment in this field.[77] The Loftuses were also prescient in foreseeing unequal distribution of access and validation for different groups of players or computer users: "Since computer literacy is becoming increasingly essential in most jobs, children who are exposed to computers early in life acquire an advantage over those who are not. The boys who outnumber the girls in the arcades will be boys who outnumber girls in the adult world of computers."[78] And they engaged with the argument against games, showing it to be a shortsighted and panicked product of intergenerational misunderstanding.

Like many admiring accounts of the early '80s, *Mind at Play* seized on examples of video games having concrete economic benefits to players. The book identifies one player in particular, a boy from Los Angeles named Mark, who was so enamored of video games that he begged his mother to buy him a computer, enrolled in computer science courses at a junior college, and used his home computer to handle accounting for a window-washing business he started.[79] This connection between games, computers, and entrepreneurial work surfaces in a number of popular accounts. *Seventeen* ran a story in July, 1983, describing a twenty-one-year-old Michigan video game player who designed games on his home computer as a teenager, sold them to a California publisher called Sirius Software, earning tens of thousands of dollars, and after several successful sales moved out west to work for Sirius full time. *Seventeen* also encouraged its female readers to submit their games to Atari and Sirius, who were "eager to accept a marketable submission from a girl."[80] An NBC News segment airing March 8, 1982, profiled Tom McWilliams, a suburban teenager from California who earned $60,000 from the proceeds of his 1981 game *Outpost*. The report framed his achievement against the initial opposition of his parents, who disapproved of his obsession with video games and refused to buy him a computer (he saved up his own earnings to purchase an Apple II). Now, the news broadcast delighted in informing us, they call him "Tommy McMillions."

The experts who saw video games as the playthings of the information revolution and a milestone in the evolution of education were no less captured by fantasy than the moral entrepreneurs stoking panic and advocating prohibition. But according to the optimistic vision, young people would be improved rather than endangered by the games. Games would be good not just for personal development, but for success at school and at cutting-edge careers in a changing world. They would help affluent families and children of the professional-managerial class reproduce their social status. To the moral entrepreneurs, video games threatened the future success of middle-class children and young adults, disrupting their education and luring them to vice. But to the academic experts, these same children stood to gain so much from playing video games that they could be understood not as diversions and amusements, not as distracting popular culture, but as essential formative experiences. To them, the new medium was bringing up a technically adept generation prepared to thrive in the computerized environment already flourishing around them.

"Donkey Kong Goes to Harvard"

The tensions between concerned local critics and intellectually engaged advocates of video games can be seen quite clearly in one moment in particular during the early 1980s when hopes and fears about the new medium were crystallized. From May 22 to 24, 1983, a symposium was staged at the Harvard Graduate School of Education on the theme of "Video Games and Human Development: A Research Agenda for the '80s."[81] The conference attracted researchers in education, criminology, psychology, psychiatry, engineering, and medicine, entrepreneurs in the educational software business, and workers from video game companies. It was funded by a grant of $40,000 from Atari, which undoubtedly invested in hopes not only of supporting academic research but also of generating positive press for the industry and countering the well-publicized position of the moral entrepreneurs. Hundreds of people attended in person, among them journalists covering the event. *Time* and *Newsweek* ran stories under the headlines "Donkey Kong Goes to Harvard" and "Video Games Zap Harvard," respectively.[82] News items in several sources conveyed mostly gee-whiz enthusiasm, with some caution from skeptics quoted for balance.[83]

Reading the conference proceedings years after the establishment of game studies as an academic field, and after the proliferation of humanities and qualitative social science research on games as communication, as art, and as popular culture, one thing that stands out is how much the focus of a games conference was on science. Participants were set on proving the value of games as advanced tools for learning, as therapeutic technologies in clinical settings of various kinds, and as wholesome rather than destructive amusements. Education experts described the utility of the video game as a "sophisticated teaching machine," capable of innovative applications in ordinary classrooms but also with patients with chronic disabilities and mental illnesses.[84] Edna Mitchell, the education researcher from the Oakland family ethnography, described how video games taught one girl to have "fast eyes," having learned to read faster by playing games, a benefit of their "cognitive workout."[85] Patricia Marks Greenfield, a cognitive psychologist from UCLA, spoke about the way the activity of games in relation to the passivity of television made games more likely to help young people develop skills. Games, she argued, teach inductive reasoning, spatial cognition, and other benefits, though she also decried the violence, racism, and sexism of many games.[86] Emanuel Donchin, a psychologist from the University of Illinois, discussed the application of games to military research and training. "From the perspective of one interested in human information processing and in the nature of interaction between humans and computers these video games provide a fantastic setting for research."[87] He argued that a computer game is a good model for understanding how a novice becomes an expert, detailing how this kind of software has been used in training research funded by DARPA (Defense Advanced Research Projects Agency) and the US Air Force.

In a session titled "Video Games and Informal Settings," many of the points concerned the everyday experience of ordinary players and the positive effects accruing to them. The main presenter was D. N. Perkins of Harvard's Graduate School of Education, whose paper title was "Educational Heaven."[88] As in the rhetoric of Asimov and the Loftuses, Perkins saw video games overcoming the age-old problem of young people's insufficient interest in learning. By being intrinsically motivating, games would lead to "strong capture" of students' minds. He referenced the "Pac-Man Theory of Motivation"—the idea that games are such a formidable stimulus to learning because of their formal features.[89] Video games are motivators by presenting clear tasks, defining the player's role and responsibilities, offering graduated levels of challenge, giving immediate and unambiguous feedback, and being a solitary experience rather than a public opportunity for ridicule or shame upon failure. Perkins cited work by another presenter, Xerox PARC researcher Thomas W. Malone, whose experimental work on intrinsic motivation was often used in support of hopeful discourses around computers and games.[90] Motivating factors according to Malone were the challenge of a clear goal with a certain outcome, the fantasy appeal of the game's scenario, and the arousal of the player's curiosity. In response to Perkins, several entrepreneurs making educational software or running computer learning companies amplified his ideas about games appealing to young people while also leading toward greater learning, whether of computers or other topics or skills. In another session on games in formal learning settings, speakers including Malone and Jerry Chafin, a special education professor at the University of Kansas, elaborated on exploiting game design for learning outcomes. Chafin sounded this particularly hopeful note: "If one could identify the motivational elements of the video arcade game and then integrate educationally relevant content into games utilizing these features, one could go a long way toward solving the learning problems of many of today's pupils."[91]

Many of the symposium speakers sought to dispel common worries, insisting that video games would promise a form of redemption through technology. This was standardly couched in terms of debate with the medium's detractors. *The Music Man* was a point of reference, and not only in the title of the keynote speech on the opening night of the conference: "Donkey Kong, Pac-Man, and the Meaning of Life: Reflections in River City."[92] Many of the speakers addressed the situation of video games delivering their benefits and promises of future benefits in the face of such deep and widespread disdain. The keynote speaker, a clinical psychologist and teacher at Harvard named Robert G. Kagan, began: "Social science feels a pressure to deliver some verdict on the wholesomeness, or lack thereof, of video games. A very immediate source of the pressure is the adults who, concerned about their children, wonder if this is not some damaging mania."[93] A brain injury rehabilitation specialist from Palo Alto, William Lynch, presented evidence that playing Atari games had positive effects on patients undergoing "remediation of cognitive and perceptual-motor deficits."[94] Games were helping them improve at "reasoning, memory, and eye-hand coordination." He also spoke out against parents and civic leaders whose criticisms "reflect incomplete understanding of both youngsters and video games."[95]

B. David Brooks, a USC instructor and consultant specializing in juvenile crime, presented a study observing 1,000 hours of arcade play by 937 young people in Los Angeles and Orange County, putting to rest the notion of games as harmful, addictive, money-draining social ills.[96] Brooks questioned whether games were addictive or isolating, whether they led to truancy, whether they turned good children into criminals and drug abusers. He argued that arcades were the new ice cream parlors, not dens of vice. Many of the children in his study did well in school and participated in other extracurricular activities. Many had consoles or computers at home. Four out of five spent less than $5 a week at the arcade, and those proficient at video games could milk one quarter for a long duration. Games themselves could even substitute for, not lead to, using drugs; they gave you a feeling of being "loaded" and anyway, being high while playing would compromise one's performance. Directly countering the panic, Brooks concluded: "Video games are a form of recreation for kids and do not really pose a serious threat to the morals of American youth."[97]

The concluding speaker at the Harvard symposium was its organizer, Inabeth Miller of the Gutman Library at the Graduate School of Education. Her rhetorical questions capture a perspective evidently shared by many of the event's participants, and more generally by intellectual and academic observers of video games in the early 1980s:

> Why are so many people entranced, held captive, by these machines? Is it visual excitement, fantasy, challenge—all the various elements described here during

the symposium? Do lights and flashing asteroids transport the participant to a place far from the boredom of everyday lives, much like watching John Travolta on the disco floor? Is escape and attraction so reprehensible that it must be stamped out, as we have seen happen in New York, New Hampshire, and Florida?[98]

Of course she did not think it should be stamped out, and she assumed that the ubiquity of video games in the home would eventually cause the panic to wither away. Actually she worried about making sure access to this emblematic technology of the new age would be equal and fair. Would computer play be yet another way of dividing haves from have-nots? Would disadvantaged children be excluded "from effectively competing in a computer-age society"? Would video games "offer only a white-oriented, middle-class picture of American life?"[99] Only a technology firmly within the grasp of America's more affluent youth, producing cultural capital for the next generation of information age leaders, could be presented in such terms.

Accounts in the national newsmagazines presented the positive emphasis of the conference's proceedings, offering tidbits of presentations on educational heaven, the production of new worlds through computer play, the therapeutic benefits of games to psychiatric patients, and the defense of children's pleasures from moral crusaders. But the balanced formula of mainstream news also prompted opportunities for skepticism and doubt. The funding of the symposium by Atari's video game money was one point along these lines, casting suspicion on the real agenda of the proceedings. Any story about video games would have to make reference to the worries of many parents about violence, wasted quarters, and the shady reputation of the arcade. For its kicker, *Newsweek* ran with this quote from Doris Mathiesen, a high school administrator from Framingham, Massachusetts, and perhaps the only doubting member of the symposium's audience: "I think Harvard has gathered all the wisest people in the kingdom to admire the emperor's new clothes."[100]

Despite the critical perspective keeping optimism in check, the mere fact that the most elite college in America played host to a conference on video games brought prestige and legitimacy to the new medium, at least opening up the question of whether video arcade critics could be overreacting. The contents of the symposium, as filtered through press accounts into popular discussions, were relentlessly upbeat. Surely many observers in the early 1980s would have been unsure of what to make of video games, which seemed at once somehow both seductive and productive, a rebellious new youth culture and good preparation for a high-tech, high-paying career.

Whatever observers made of them, seeing elements of good or bad in their potential, the framing of the new medium through many types of public discourse attributed great power to these amusements to chart the younger generation's course through later life.

Video Man, *WarGames*, and Techno-Ambivalence

While their status as harmful or educational was a theme of news reports and columns in papers and magazines, a more fantastical and symbolic register of concern could be found in fictional stories about video games in cinema and television. Several movies released in the 1980s have promi- nent video game themes, and one motif they tend to share is crossing through the boundary separating the real world and the game. This notion of the player entering into a diegesis is similar to fiction/reality play in movies like Buster Keaton's *Sherlock, Jr.* (1924) and Woody Allen's *The Purple Rose of Cairo* (1985): characters pass from one realm to another and then return. Dreamlike fantasy is the essence of this device; the game world is an idealized, highly dramatic scenario. For video game narratives, the fantasy is to be doing more than merely controlling patterns of light and sound, a fantasy of identification with the representation in the game. It's all about the boy's or man's mastery and power.

This trope could also be an expression of one common fear around arcades and computers: of the child getting lost, of the isolation of play capturing youth who are no longer present in real space and real life. The illustration on the cover of the 1982 *Time* magazine issue announcing "Video Games Are Blitzing the World," which we saw in chapter 1 (fig. 1.1), pictures a male figure with a gun inside the representation in an arcade cabinet screen. This kind of imagery would dovetail with widely circulating ideas about the games being mesmerizing and potent, exerting a force over their user. Films working along these lines include *Tron* (1982), *WarGames* (1983), *The Last Starfighter* (1984), and *Cloak and Dagger* (1984). They pres- ent young male protagonists finding escape or empowerment through technology and adventure. As Carly Kocurek describes them, the heroes of *Tron* and *WarGames* are "exceptional embodiments of the technomascu- line," the latest versions of an American pop cultural type, "the bright, capable, mischievous, tech-savvy boy."[101]

An early '80s Saturday morning cartoon television series, *Spider-Man and His Amazing Friends* (NBC, 1981–83) and the Hollywood film *WarGames* both present revealing variations on this fiction/reality trope. The *Spider- Man* episodes do involve the fiction/reality transgression similar to *Sherlock,*

Jr. or *Tron*, which presents a young man who is enveloped in the computerized environment. *WarGames* works slightly differently, by the player stumbling upon a networked computer program that he believes to be just a game but which turns out to be a real military program. In both *Spider-Man* and *WarGames*, we find many of the same hopes and fears expressed in contemporaneous news media. In these two examples, the optimism and pessimism around electronic games are married in a way that shows how, rather than either hopeful or fearful, we might best see ambivalence about technology as the prevailing discourse of video games at the point when they became a pop culture phenomenon.

Spider-Man and His Amazing Friends was an animated series on NBC that ran a total of twenty-four episodes in its original run (and kept airing repeats for years after that). Following the standard format, the superhero friends Spider-Man, Iceman, and Firestar have alter egos as students at Empire State University by day but also transform into crime-fighting heroes to do battle against supervillains. A character introduced in 1981 in the second episode, Video Man, appears three times in the series: first for two episodes as a villain, and then a final time as a hero. Video Man first appears as a two-dimensional, pixelated, hulking foe made to resemble a video game sprite. In his first episode, he is a depersonalized villain but the association with video games makes him seem like a computerized enemy who is more machine than man. At one point he lets video game characters loose from their game cabinets at Earl's Arcade to wreak havoc, and a Pac-Man-esque character, "Mr. Grabber," munches through a park with his two sharp teeth in pursuit of Spider-Man.

In the series' seventh episode, "Video Man," Peter Parker and his friends are shown hanging out like regular teenagers at Earl's Arcade. From his command center, the supervillain Electro controls the machines in the arcade, and Electro sends Video Man through the electrical lines and arcade cabinets to capture an all-American hunky boy, Flash, a friend of the heroes, who is such a whiz at video games that they distract him from his university studies. Flash is captured by the villains and taken to a physics lab, where he is caught inside a *Pong*-style game in which he dodges balls bouncing back and forth. When Flash is hit by the one-hundredth ball, Electro will kill him. Eventually Spider-Man's superhero friends Iceman and Firestar are also trapped inside the villains' games, one that resembles *Asteroids* and another that is a vector graphics auto racing game. Somehow they figure out how to enter Flash's *Pong* game to defend their friend, and Spidey gets Electro and Video Man to destroy each other. Having vanquished the villains, the heroes return safely from dangerous game space to reality.

Flash's memory of the episode is supposed to be wiped clean, but in a (perhaps ironically) moralizing epilogue, he has a flashback of being tortured by Electro triggered by a seemingly benign game of *"Pongle."* The episode concludes on Flash's lesson learned: *"Pongle!* I can play anything but *Pongle.* Excuse me I gotta go to class, lift some weights, run some errands, I gotta get outta here!" He flees the arcade. Angelica (alter ego of Firestar) winks at Peter (Spider-Man) and Bobby (Iceman), and the episode ends on this wholesome turn away from the seductive dangers of electronic play, with the All-American, middle-class, white teenage boy scared straight.

Video Man returns in season 3 of *Spider-Man and His Amazing Friends* as a good guy in the episode "The Education of a Superhero" (1983). Gamesman is a new video game villain, a maniacal genius preying on the kids at the arcade. He uses the arcade games for mind control of the young players. One nerdy kid, Francis, is better than all others at one game in particular, *Zellman Command*, and when Gamesman explodes the *Zellman* cabinet, Francis is sucked into the game to become a superhero, Video Man. Video Man can move through electrical lines in and out of television sets and arcade machines. Gamesman threatens to use telecommunications technology (TV sets, satellite transmission) to exert mind control over the whole world. But of course Spider-Man and Video Man foil this plan and save all of humanity from an awful fate.

Putting aside the minutiae of the convoluted plots in these cartoon episodes, the transformation of Video Man from a villain using video games to cause harm to young players into a hero whose superpowers arise from proficiency at an arcade game (and counteract the power of broadcasting as mass media) speaks to the uncertainty over the value of the new medium. The show at once satirizes and trivializes outcry over games' effects and reinforces commonplace fears about them. But by shifting the character from evil to good, it also shows a progression from fear of games as a moral hazard to appreciation of games as advanced technology involving complex skills that kids can learn to master, gaining a valuable advantage.

WarGames is similarly the story of a teenage boy of the video game generation, David Lightman (Matthew Broderick). He frequents a typical arcade in his Seattle neighborhood where he plays *Galaga*, the colorful post–*Space Invaders* alien invasion shooter. He evidently prefers arcade games to schoolwork, as he races from the Grand Palace arcade to the high school, arriving late and receiving a big red "F" on his biology test. David is also eager to play games after school using a home computer in his bedroom connected by modem and telephone line to a network. After showing off his hacking skills by gaining access to his school's computerized records and changing his failing biology grade to avoid summer school, he begins to play a game he discovers called "Global Thermonuclear War." David assumes this is just a simulation, but it is no such thing. It is really a poorly secured NORAD military program for controlling a nuclear conflict with the Soviet Union, capable of launching missiles at the enemy when under attack without human input. A Cold War nightmare ensues as an accidental nuclear crisis is precipitated by an unsuspecting thrill-seeking teenager. David is both the cause of the crisis and the one to avert World War III through his bold and fearless initiative, computing knowledge, and natural intelligence. (He seems the type not to let school interfere with his education.) Along with the audience, David recognizes that nuclear war means mutually assured destruction, a pat Cold War moral made more appealingly contemporary with the home computer and video game twist.

While perhaps richer in humanist antinuclear and anticomputerization themes, *WarGames* also expresses the unexamined assumption that video games are an affluent teenage boy's property, having the potential both for personal intellectual enrichment and extreme mischief and danger. David's adept hacking produces much more serious consequences than the fun he is looking for. The "mere play" of electronic games matters more than anyone could imagine. The crisis he unwittingly triggers is outlandish and implausible, but the notion of video games being hazardous was everywhere in 1983. It's a Hollywood premise that video games could cause World War III, but this was also a hyperbolic expression of commonly held views about the new medium. Champions and detractors alike agreed that games were sure to produce significant effects, in particular on well-off teenage boys like David Lightman. His friend and eventual romantic interest Jennifer (Ally Sheedy) shares in the adventure but has little interest in or proficiency with technology. This gendered pairing of the hero and sidekick places video games and computers squarely in the boy's domain. In the climactic scene at the NORAD Command Center, it is David's familiarity with computers and games that helps him find the solution to mutually assured destruction: he teaches the computer controlling the war game about futility by having it play a series of tic-tac-toe games to a draw, and the computer learns that, as in tic-tac-toe, in global thermonuclear war, "the only winning move is not to play." This is the solution that saves the planet from annihilation, from the extinction of humanity. David's knowledge of cutting-edge technology gets him into the conflict that gives *WarGames* its dramatic interest, but it would certainly also prepare him well for work in the information society if it doesn't kill him, and the rest of us, first.

Stories of young male video game players in early 1980s pop culture exploit the novelty of computers in everyday life and of video arcades defining a generation. They also share some of the same ideas as *The Music Man* and a longer history of narratives about young people's leisure. The hysterical excesses of danger in the melodramatic plots of *Spider-Man* and *WarGames* are also representations of the never-ending struggle to "keep the young ones moral after school." Also like *The Music Man*, video game stories are hardly endorsements of media panic encouraging unwarranted concern. Never mind that these are films or TV shows aimed at children. They express an appropriately complicated, conflicted stance on new media and the generation coming of age along with its emergence, acknowledging its disruptive energy but also seizing on its promise of a better tomorrow.

6   Pac-Man Fever

I got Pac-Man Fever
I'm going out of my mind.
—Buckner & Garcia, "Pac-Man Fever"[1]

Pac-Man Fever was the Beatlemania of Generation X. The first video game to fully permeate North American popular culture, *Pac-Man* stood for the medium itself at the moment when video games became enshrined as a mainstream form of leisure. In the early 1980s, *Pac-Man* was the most successful game commercially and the most popular game with the widest range of players: boys and girls and children and adults, though kids were the main consumers of games like *Pac-Man*. It inspired an enormous array of spinoffs, adaptations, and merchandise bearing the iconic yellow circle with a missing wedge and a black dot for an eye. It is impossible to imagine the emergence of the medium without *Pac-Man*. Along with a handful of other games like *Pong*, *Space Invaders*, *Tetris*, and *Angry Birds*, *Pac-Man* is a game that everyone seems to know and appreciate—a classic. Sherry Turkle called it "the first game to be acknowledged as part of the national culture."[2] *Pac-Man* is like a perfect pop song: it seems so simple, so catchy, so universal in its appeal. No game more strongly defines the first decade of the medium.

This is remarkable for several reasons. *Pac-Man* was unlike most games in its aesthetics and representation. It has a simple interface—just the four-way directional joystick—and a novel concept in which you navigate a maze, eating dots or monsters to collect points and graduate to the next level. The images of the yellow Pac-Man and brightly colored ghosts were light-hearted at a time when most games had imagery deriving from science fiction or war. Many of the popular games at the moment of *Pac-Man*'s initial popularity were about shooting or driving (in some games like *Tron*, you would get to do both) and had militarized space adventure scenarios.

In distinction to the electronic zapping, blasting, bombing warfare noises of games like *Asteroids, Galaxian, Tempest,* and *Zaxxon,* a game of *Pac-Man* starts off with a cheerful, funky four-bar intro in a major key. As you move around the maze, the game plays an uptempo waka-waka rhythm to accompany the yellow figure's munching. The Pac-Man is animated to open and close his mouth perpetually, a ravenous sprite. When you lose in *Pac-Man,* the music is a descending glissando pitying you upon your demise as the yellow figure vanishes. The game is an amusing cartoon, not an intense space opera. Pac-Man is a friendly character of the kind found in comic books and Saturday morning television for young children, not a spaceship in battle over the fate of the universe. The *Pac-Man* arcade cabinet art pictures a rotund comic book figure with enormous eyes and frog feet. It's goofy, not tough or adventuresome. The hit spinoff *Ms. Pac-Man* (1982) interpolates snippets of romantic comedy between its levels, showing Pac-Man and Ms. Pac-Man progressing from their "meet cute" to reproduction.

The most notable commercial success in video games before *Pac-Man* was *Space Invaders. Space Invaders* was the first massively popular shooter game. It picks up on several traditions of representation and amusement: sci-fi alien invasion narratives like *War of the Worlds* (it was inspired by the 1953 film adaptation) and electro-mechanical shooting gallery rifle games from the penny arcades and sportlands. In *Space Invaders,* you are defending civilization from assault by killing the enemy. It was no coincidence that *Space Invaders'* popularity coincided with the dominant presence in popular culture of *Star Wars,* in which warfare in a distant galaxy plays into boy-culture ideals of courage in battle against a mortal enemy. Many other versions of this scenario were successful as video games, including *Missile Command,* with its Cold War dystopia of nuclear conflict, and *Galaxian* and *Galaga,* which update *Space Invaders* with color graphics and other technical innovations. *Asteroids, Defender, Tempest,* and many other space battle games worked in similar ways by tapping into masculine fantasies of power and heroism.

Like any video game of this period worthy of sustained attention, *Pac-Man* had considerable complexity and scaffolding levels of difficulty.[3] To master it was no easy feat, and its challenge inspired competitive, intensive play among the denizens of the arcade and, eventually, home video game players. *Pac-Man* is a chase game with a twist: you, as the Pac-Man, are evading your enemies, the brightly colored ghosts Blinky (red), Pinky (pink), Inky (cyan), and Clyde (orange), while trying to eat all 240 of the pellets in the maze (collecting points as you consume them), until you

Figure 6.1
Pac-Man cabinet with cartoonish characters.

swallow one of four larger power-pellets also known as energizers, one in each corner of the maze, which effects a dramatic reversal. All of the ghosts turn royal blue and are transformed from lethal to vulnerable. The monsters reverse direction while you attempt to chase and consume them before they revert to their usual red, pink, cyan, or orange. For this brief time, the hero and the enemies trade roles of prey and predator. But if you try to eat a blue ghost at the moment it is changing back, you die, so one critical element of the game is timing your attack and knowing when to retreat.

Pac-Man's gameplay involves evasion and pursuit, a race against time, and navigation of a maze, often applying a strategic pattern to guide your route. Each of the four ghosts is programmed to move in its own pattern, which a skilled player can come to appreciate and predict. (Many guide-books offered instruction on maze pattern strategy, and players standing around *Pac-Man* cabinets learned patterns from one another.) After clearing a board of all the dots big and small, a fresh maze appears and the challenge increases. The boards are named for a colorful fruit (and on later levels, a galaxian, bell, or key) that appears periodically in the center, which can be eaten for points—the later levels have more valuable fruit. On more advanced levels, Pac-Man moves faster; on even more advanced levels he slows back down, though the enemy ghosts do not. On higher levels, ghosts are blue for a shorter time after you eat a power-pellet, making it harder to eat them. Eventually they stop being blue at all. At 10,000 points you earn an extra man, but if you die three times (the extra is a fourth) your game is over and you have to drop another quarter in the slot.

    Pac-Man works exceptionally well as a game, rewarding both casual attempts and devoted, repeated play. It is fun on the first quarter and the hundredth.[4] When you die, you are likely to want to try again. It has many appeals: colorful characters and fruit, humorous sounds, an easy interface, a novel concept (at least in the early '80s), and the serious challenge of mastering its increasingly difficult levels. It is not surprising that *Pac-Man* became popular, but it is quite notable that the one game that reached its unparalleled status was unlike most video games in some important ways. *Pac-Man* and its various spinoffs and adaptations were never as masculinized as *Space Invaders* and those games influenced by it. *Pac-Man* had a more inclusive, egalitarian quality that opened video games to more players and softened the medium's reputation as aggressive and violent. *Pac-Man* was instrumental in making video games a mass phenomenon, and its value to the games industry and also to the culture of electronic play was in its expansion of the market beyond the young and male crowd. Video games became mainstream entertainment and a pop culture craze at the moment when *Pac-Man* made them seem safe and inviting for all.

The Woman's Video Game

When *Pac-Man* was released in 1980 by Namco, a Japanese firm, video games had become well established as a profitable industry and a popular form of public and private amusement, but it was clear to all involved in the video games trade that young and male players were their main customers. *Pac-Man* was designed to increase the market size for video games by appealing to women in particular, drawing them into the game rooms that had in some ways seemed forbidding to female players or to opposite-sex couples. Appealing to women had several virtues for game manufacturers, operators, and arcade proprietors. It not only expanded their customer base, but also brought a measure of respectability to the business and medium of video games. A greater gender balance in arcades and other venues of public coin-op play would have the potential to change their reputation. Women or teenage girls were also regarded as a potential magnet for male customers, under the assumption that boys would follow where the girls are. While it may not have been anyone's intention to make a more child-friendly game, *Pac-Man* also had the virtue of being cute and cartoonish rather than aggressive and violent just at the time when games were a cause for concern because of their focus on shooting and killing enemies. *Pac-Man* would seem less likely than the space and war games to be corrupting the youth.

Toru Iwatani, the young designer who created *Pac-Man* for Namco, has spoken repeatedly of his inspiration in making a game to appeal to women. The popular games at the time were all about shooting aliens, which was not perceived to be a theme that appealed to female players. In order to get women and couples into game rooms, Iwatani designed *Pac-Man* to have a specifically feminine draw: "When you think about things women like, you think about fashion, or fortune-telling, or food or dating boyfriends. So I decided to theme the game around 'eating'—after eating dinner, women like to have dessert." He has often compared the form of Pac-Man to a pie with a missing wedge: "If you take a pizza and remove one piece, it looks like a mouth. That's where my idea came from." The ghosts were inspired not by the sci-fi or space opera tropes of most popular games, but from *anime* and *manga* along with American children's comics and cartoons. The ghosts were like Casper or Obake no Q-Taro. Powering up the Pac-Man by eating an energizer pill was inspired by Popeye eating his spinach to gain the strength to defeat his foe Bluto. In Japan, the game was not initially terribly popular, but it appealed, according to Iwatani, to "people who didn't play games on a daily basis—women, children, the elderly."[5] As Tristan Donovan observes in his history of video games, *Pac-Man* draws on the Japanese concept of *kawaii*, often translated as cuteness, typified in global popular culture by Hello Kitty.[6] Many observers both in the '80s and later on identified cuteness as one of *Pac-Man*'s key appeals, and cuteness has never been aligned with the dominant sensibility of teenage and young adult masculinity in the United States.[7]

The name *Pac-Man* comes from the Japanese *puku-puku*, an onomatopoeia that references the sound of a mouth opening and closing. It is named for the act of incessant eating. Originally in Japan the game was named *Puck-Man* but when Midway licensed the game to release it in the United States, it changed the name to *Pac-Man* to avoid any vulgar vandalism. Upon its release stateside in October 1980, the game exceeded expectations. In Japan, it had been less popular than the contemporary Namco game *Galaxian*, the variation on *Space Invaders*. But in North America, *Pac-Man* appealed widely to children and women in addition to male players, and to both more casual and more serious visitors to arcades.[8] Its financial impact was staggering: it earned $1 billion in revenue in its first fifteen months in the United States, and the home console version was predicted in the pages of *Time* to be a bigger money-maker than *Star Wars*, a widely recycled factoid.[9] America had almost 100,000 *Pac-Man* arcade cabinets collecting quarters in 1981, and soon enough more than two million *Pac-Man* game cartridges would be sold to consumers with Atari consoles at home.[10] The release of this ultimately disappointing product, which sold well but was widely regarded as a pale imitation of the arcade version, was promoted by a massive marketing stunt called National Pac-Man Day on April 3, 1982, with public events staged at malls and other public places in many American cities.[11] *Pac-Man* merchandise from apparel and lunchboxes to wristwatches and drinking glasses was for sale in every department store and souvenir shop. The 1982 World's Fair, in Knoxville, Tennessee, had not only several game rooms filled with video arcade cabinets, but four Pac-Man souvenir shops selling shirts, mugs, pins, posters, pillows, stickers, Pac-Man Putty, miniature souvenir license plates, tote bags, and a child-size sleeping bag.[12] There were various *Pac-Man* spinoffs and adaptations, including a tabletop Coleco game, a Bally pinball game, and most famously the feminized version of the game hailing the interest of girls and women, *Ms. Pac-Man*, which debuted in January 1982 and became hugely profitable in its own right. *Pac-Man* and *Ms. Pac-Man*'s unprecedented success were routinely attributed to their ability to get women to overcome whatever reluctance or hostility they had to playing coin-operated video games.

Popular press stories often highlighted the female participation invited by *Pac-Man*. An NBC News report on May 25, 1982, on Pac-Man Fever noted not only the game's outstanding commercial fortunes, but also its unusual demographic appeal. An interview with Michael Blecha, a "Pac-Man Retailer" framed against a wall of Atari cartridges, offered this observation: "Women are insane about this game. Men like the sports games, the action games and the space games. Women like the predator games." The report continued by showing images of a store selling *Pac-Man* products including T-shirts, coffee mugs, beer mugs, tote bags, pillows, balloons, hats, board games, books, dolls, hand puppets, and neckties. *Pac-Man* was a mass market bonanza, a Mickey Mouse for the electronic age. The report concluded with a shot of hands counting money. In the pages of the November 1982 issue of *Working Woman* magazine, the success of *Pac-Man* was attributed to "the huge numbers of women who play." In drawing the familiar distinction between *Pac-Man* and war games, the author framed video games pre-*Pac-Man* as an exclusionary medium in which men played and women watched. *Pac-Man* liberated women from their place by the sidelines. *Working Woman* explained the appeal of *Pac-Man* as one of genre distinction: "Video experts believe that Pac-Man's lighthearted graphics, catchy tunes and the absence of exploding space-ships attract women players."[13]

A first-person column in *New York* magazine in 1983 by Jennifer Allen chronicled her descent into *Pac-Man* addiction, a fix she satisfied at a coin laundry on the Upper West Side of Manhattan where other players around the machines were "skinny and black and ... about fourteen years old." By contrast, she felt "like the chaperone at the party." *Pac-Man* became her game upon her first encounter with it at a Catskills resort game room. In contrast to the players at the space/shooting games, the ones who preferred *Pac-Man*, adults and children alike, were "laughing and talking, better tempered than the dour, determined players who wrestled with Asteroids and Space Invaders." This more sociable interest in play would put distance between *Pac-Man* and the more aggressive and violent arcade games, which the writer implied endeared *Pac-Man* to female players. "There were no explosions or smashups; when a Pac-Man got eaten, the only sound was a droopy, wilting noise, the kind that might accompany a clown making a sad face."[14] The news interest of this story, which appeared in 1983, would have been not just the topicality of the video games craze and of Pac-Man Fever as an emblem of all that, but also the man-bites-dog quality of a grown woman being a serious player of video games.

All of this newfound feminine appeal occurred just at the moment of media panic around video arcades, just as moral authorities and guardians of virtue agitated for regulation of game rooms and video game cabinets in the name of protecting youth from this fresh danger. One reason why *Pac-Man* and its many spinoffs were so welcome in game rooms, stores, and the many other spaces of play was that girls and women were regarded so favorably by proprietors of these venues. Unlike male patrons, girls and women were not often seen as corrupting or threatening influences. On the contrary, their presence was reassuring of the wholesomeness of a space and of the activities going on within. According to one study of young people's attitudes about games and arcades, some female patrons worried that their presence in game rooms filled with pinball and video machines was somehow improper, but changing the image of games could change this perception. The campaign in the coin-op trade to increase female patronage was good business not just in the sense of attracting more customers, but also in the sense of remaking the image of coin-operated amusements as legitimate and culturally worthwhile.[15]

By appealing too much to "loners" and "losers," the coin-op trade risked its reputation. In a polemical column in a January 1982 issue of the coin-op trade paper *PlayMeter*, Marion Cutler and Jane Petersson argued that this reputation was critical to the fortunes of the whole industry, and that "distaff fans" of video games were crucial to turning things around: "Women—more than zaps, lights, or electronics—are the signal that the game has become respectable." They made an impassioned case that the taverns, arcades, and other spaces of electronic play had been inhospitable to female patrons not just because they were shady or dusty or unsafe, but also because the games themselves were a big turn-off: "With half-naked female figures and gigantic 'zooms' and 'zaps' forming the backgrounds, pinball and electronic games and their surroundings fairly shriek 'Men Only!'"[16]

The imperative to get women into arcades could be accomplished by recognizing their newfound social and economic mobility in the wake of second-wave feminism. The coin-op trade could capitalize on women's expectations of equality and their increased spending power by offering experiences of leisure and play hospitable and desirable to them. If women expected to be treated as equals and to have opportunities to enter any realm open to men, they would naturally find their place in arcades and game rooms. But this might necessitate catering to them to win this business. It might be particularly appealing, Cutler and Petersson argued, if the spaces of play could become sites of courtship—pickup or date spots. The appealing scenario they offered *PlayMeter* readers, typically men in the amusements trade, was that upwardly mobile women would begin to patronize their businesses as not only an expression of their equality with men, but also as a way of meeting eligible bachelors. In the early 1980s, when girls and women were discussed as potential patrons spending money on video games, ideas about romance and courtship were often part of the story.

The way to get women into arcades was most often by offering the kinds of games they were believed to prefer, and games with cute or cartoony themes were most closely identified with female players. In addition to *Pac-Man*, this category would include *Donkey Kong*, *Frogger*, and *Centipede* (a rare early game to have been designed by a woman), though *Centipede* was also a shooting game in the *Space Invaders* style. In explaining "The Video Games Women Play … And Why" in the May 1, 1982, issue of *PlayMeter*, Mary Claire Blakeman made clear not only that such "games where you can recognize things" were a positive appeal, but also that games of "the space wars genre" were a turnoff because of their association going back several decades with astronauts and science-fiction movie heroes who were always male. A businesswoman in Oakland offered that she prefers "games where you can enjoy the graphics even if you lose." On the other hand, she did not favor "the space games where things are coming at you out of nowhere." *Pac-Man* was singled out as a particular favorite, and the author's identity as a woman was front and center in this perhaps tongue-in-cheek evaluation:

> I don't care what the socio-political experts say—I like *Pac-Man* because it's the only game where you can eat your way out of your troubles. Not only that, but the more you eat the better you are. What a nice antidote to all those diet books, aerobic dance classes, and medflies in the pantry. In *Pac-Man* it's simply—the more you eat, the more you win.[17]

But beyond its content, *Pac-Man* was also celebrated for its simplicity and ease of play, and for its sense of fun. It demanded less dexterity and devotion than some games, which made it more inviting for casual players. *Pac-Man* was also a game well suited to courtship by being less gender specific and masculine, more neutral. A boy might like to show a girl how to play, or to impress her by achieving a high score similar to his feats at a carnival game winning her a prize.[18]

In the publications aimed at fans of video games, which exploded in number in 1982 and 1983, we find not only the assumption that video games were especially for a subculture of boys and young men, who are the implied reader except in rare instances of broader address, but also that *Pac-Man* went against the grain, becoming more popular than any other game through its mass popularity, which exceeded subcultural bounds. One of several guidebooks published at this time with the title *How to Win at Video Games* rated a number of popular arcade games by their "sex appeal" (meaning the ratio of male to female players). *Defender* was rated 95 percent male. *Asteroids*, "stereotypically, a male game," was rated 60 percent male. *Donkey Kong* and *Centipede* were both rated 50–50 male–female, and of the latter the author wrote, "Cuteness overcomes most people's squeamishness about bugs." Only *Pac-Man* was given a ratio favoring female players, 60–40, with a note that its "strong appeal to women [is] accredited to its non-violent theme."[19]

Often these books and magazines tried to explain *Pac-Man's* amazing ubiquity, its status as the number one game by far. "One of the big reasons for Pac-Man's popularity," asserted another 1982 volume simply called *Video Games,*

> is that girls like him just as much as boys. Before Pac-Man, arcades were mostly for boys who liked to play the space zap games. A lot of people thought that would never change. Even arcade owners were surprised when girls started show-ing up to play Pac-Man. Of course, girls' interest in video games wasn't caused just by Pac-Man. Video games had spread to restaurants, airports, supermarkets, convenience stores—places that weren't strictly boys' territory—and the girls started to play them.[20]

The popularity of *Pac-Man* was also explained by reference to the game's simplicity and cuteness and its "very human, very personal" graphics.[21]

The bimonthly magazine *Vidiot,* published by *Creem* magazine in 1982 and 1983, celebrated Pac-Man as the first "computer-generated pop star" by emphasizing his cuteness and simplicity as the way to economic good fortune for the games industry. Pac-Man was "the first electro-terrestrial with personality," "the Beatles of his day."[22] But *Vidiot* also referenced *Pac-Man's* amazing popularity as a product of crass hypercommercialism, which it surely was. A headline on the February–March issue of 1983 screamed "Pac-Man Sells Out!" The unflattering association between con-sumer culture and femininity fit with *Vidiot's* convergent topics of rock music and video games, with its frequent photos showing mostly male rock stars posing by video games with captions like "The Rockets are Vidi-ots!" In a roundup of *Pac-Man* merchandising, which is supposed to have earned $10 million "this year," the magazine presents snarky put-downs of products for sale in stores including fruit-scented erasers, ball darts, rearview mirror hang-ups (ghosts in place of fuzzy dice), punch balls, pil-lows, shower curtains, things you stick on the end of a pen, bumper stick-ers, board games, thermal underwear, lunchboxes, puppets, and shoelaces. The final entry describes a cigarette lighter imprinted with the *Pac-Man* logo, for which the author recommends the following use: "Take all the Pac-Man merchandise you can find, arrange it neatly in a heap and apply this product."[23] In a box alongside this write-up ran a review of the new ABC TV series from Hanna-Barbera, referenced as *The Pac-Man Cartoon Show,* treated as a curiosity but a dull and predictable one.[24] The popularity of *Pac-Man* might have demonstrated that video games were now part of the national culture, but it also came at the expense of the subcultural cohesiveness of serious video game players (the type implied in the address of fan publications like *Vidiot*). Their defensiveness about the commercialism around *Pac-Man* was not expressed in overtly gendered terms, but the contrast between a young, male subcultural identification and this mass market schlock is an undercurrent in fan discourse, especially considering how regularly *Pac-Man*'s identity was as the one video game women really like.

This identity is underlined in another ironic fan publication, Mark Baker's satirical book *I Hate Vidiots*, published in 1982. In a combination of the style of *The Official Preppy Handbook* (1980), with its brief sections and lists and drawn illustrations, and the irreverent satire of *Mad* magazine, *I Hate Vidiots* skewered video game culture at the moment of its ascent. It travestied the players of video games by including a chapter titled "Psychological Profiles of the Vidiot by Game," using the male pronoun in its descriptions of each game's player profile. Of *Missile Command*'s player, Baker writes: "He looks at life through the window of vulnerability."[25] Of the *Defender* player: "This vidiot would seem to have a lot of things going for him: manual dexterity, mental agility, and especially a self-sacrificing nature."[26] And for the player of *Tempest*: "If he weren't addicted to *Tempest*, he'd probably be an acid-head anyway."[27]

But *Pac-Man* is different, and the appeal to women is taken here as an opportunity to sexualize video games, and both the male and female players who enjoy them, in terms of gendered desire and pleasure. This sexualizing rhetoric follows many representations of girls or women and video games, such as the assumption that the presence of girls means that arcades become sites of courtship, or the frequent inclusion of shots of women in fan magazines like *Vidiot* modeling T-shirts or promoting subscriptions, similar to the use of posed models by motorcycles and hot rods. So in *I Hate Vidiots*, the appeal of *Pac-Man* to female players becomes an opportunity to stereotype sexual difference:

> Pac-Man is a video addiction preferred by female Vidiots (so right away, you know there is something funny about a guy who plays this game). Women don't like video games because they're all about little guns and missiles, shooting, firing fingers on buttons, thrusting, battering, penetrating, explosions throbbing, pulsating, the earth moving in a final shudder of … Ahem, well, girls don't really like them.
>
> Pac-Man is different. Pac-Man provides a thrill that the female Vidiot can enjoy. It's all about playing coy, running, hiding, waiting until the little guys

almost get you cornered, then turning suddenly to pounce like Cat Woman, eating, engulfing, swallowing, scratching, clawing, churning, engorging, devouring, throbbing, pulsating, the earth moving in a final shudder of ... Whew! It is getting hot in here, or is it just me?

Pac-Maniacs are Vidiots who never get called for dates. They just go down to the arcades to snicker at the dopey guys down there. They use their reflection in the screen to put on their makeup. And they'll never get a date either, if they don't stop hanging around in the corner of the arcade giggling with the other girl Vidiots.

Famous Pac-Man Vidiots: Britt Ekland, Linda Lovelace, Tennessee Williams, Truman Capote.[28]

By including two female actresses, one a Swedish movie star and sex symbol and the other famous for appearing in *Deep Throat* and other porn films, along with two famous gay male writers, as the "Pac-Man Vidiots," Baker underlined and stigmatized the sexual difference meant to characterize the one video game not identified with straight, young, male players. The illustration accompanying this psychological profile was of an attractive woman in repose on a bed under a blanket pulled up to cover her chest, with a *Pac-Man* arcade cabinet in the bed beside her, and both of them smoking postcoital cigarettes.

Figure 6.2
Illustration from Martin Barker, *I Hate Vidiots*, sexualizing *Pac-Man* and its female players.

Ms. Pac-Man

This emphasis on eroticizing *Pac-Man* and its players was carried along into the release of the game's most successful arcade sequel, *Ms. Pac-Man*, in 1982. Many copycat games attempted to capitalize on the munching maze concept, including *K. C. Munchkin*, a game cartridge for the Odyssey[2] console removed from the market in 1982 after Magnavox lost a copyright infringement case.[29] *Ms. Pac-Man*'s origins were in Bally Midway's efforts to release a homegrown version of its own smash success as the company licensing Namco's import. *Crazy Otto*, a copycat maze game created by an American company called General Computer Corporation (GCC), became *Ms. Pac-Man* when Bally Midway acquired it and changed the name and some small details of character design. *Crazy Otto/Ms. Pac-Man* was in some ways an improvement on *Pac-Man*. While the basic design—the maze, the dots, the ghosts, the joystick—remained the same, the newer game had four different maze configurations of increasing difficulty, and the patterning of ghost movement was more challenging. In place of stationary fruit, the new game's fruit floated through the maze, adding the challenge of another element of pursuit. *Pac-Man* and *Ms. Pac-Man* were quite similar, but the latter seemed like an evolutionary step forward, and its popularity extended and amplified the influence of *Pac-Man*.

The animations in between levels of *Ms. Pac-Man* also subtly built on the branding of the franchise. Even as *Crazy Otto*, the game had animations functioning as intermissions between mazes (*Pac-Man* had these too), each one introduced with a clapperboard conveying that these are little cinematic narratives. Originally these intermissions used the sprites of the *Crazy Otto* characters but when the game was made over as *Ms. Pac-Man*, they were changed to suit the theme. In the first movie, "Act 1: They Meet," following the second board, the courtship of Pac-Man and Ms. Pac-Man occurs during their pursuit by two different ghosts. The hero and heroine find each other during their comical evasion of the bad guys, and at the end they run off together. After the fifth board comes "Act 2: The Chase," against an electronic rendition of a cheerful ragtime tune suitable for silent movie accompaniment. This movie is an homage to the chase scenes of early cinema. Alternately the girl pursues the boy across the screen and vice versa; their chases speed up, and the brief segment ends without resolution, a cliffhanger. Finally comes "Act 3: Junior" after the player has cleared seven boards. Pac-Man and Ms. Pac-Man are pictured together at the bottom of the screen as a stork flies in and drops a bundle that lands by the two characters: inside is a baby Pac-Man. According to Doug Macrae, who worked on the game for GCC, the name *Ms. Pac-Man* was chosen over *Miss* or *Mrs.* not as a gesture toward women's liberation but rather to avoid any suggestion that the offspring arriving in Act 3 of the game's intermission movies has been born out of wedlock.[30] In the cartoon adaptation on TV, *Pac-Man*, which ran for two seasons between 1982 and 1983, the Pac family includes a wife, Pepper, and a baby. Pepper has a stereotypical female role as wife and mother rather than as protagonist leading the efforts against the evil ghosts. The family scenario carries over from the arcade game to this spinoff.

The decision for Bally Midway's sequel to have a female character was seen straightforwardly as an effort to capitalize on the popularity of *Pac-Man* with female players. *Electronic Games* magazine told fans that "Midway wanted to thank the scores of female arcaders who took Paccy to their hearts by producing a female version of the coin-op classic."[31] Bally/Midway promoted *Ms. Pac-Man* as the "femme fatale of the game world," and the character and arcade cabinet design emphasized femininity while maintaining the cute and cartoonish style of the original. In a Bally Midway flyer (an advertisement aimed at arcade game operators), a *Ms. Pac-Man* is played by a teenage girl in jeans and sneakers with long blonde hair while a boy looks on with excitement. On either side, grown men play video games with space/shooting themes: a *Bosconian* and a *Galaga.*[32]

On the screen, Ms. Pac-Man looks the same as Pac-Man except for her red lips, red hair ribbon, and the addition of an eye and a beauty mark. (Pac-Man was often represented with a black dot for an eye, but not in the arcade game itself.) On the cabinet and marquee, *Ms. Pac-Man* is certainly girly. On the marquee (the glass-covered signage above the screen), yellow type is outlined in pink and all of the graphic elements are set within a yellow-outlined pink rectangle with rounded corners against a light blue background. Ms. Pac-Man wears blue pumps and heavy makeup: big pink lips pursed as if to kiss, brightly blushed cheeks, long lashes, and eyelids covered with blue eye shadow. She sits in the valley formed atop the letter M in a glamor pinup pose, one hand behind her head and both knees raised. The menacing, masculinized ghost is pink too and he leers at her. The side of the cabinet, less often seen when arcades position games lined up along a wall, repeats the same color and graphical motifs, but including the ghost's pursuit of Ms. Pac-Man, who is running away on her high heels while now wearing large pink gloves.

Figure 6.3
1982 Bally/Midway flyer showing *Ms. Pac-Man* and its intended market.

Figure 6.4
Ms. Pac-Man marquee with its feminized representation of the character and the game.

Many publications picked up on the female audience intended for *Ms. Pac-Man* and its effort to capitalize on the broadening of video games' demographics that *Pac-Man* represented. The female version was often referenced in highly gendered terms in a way that *Pac-Man* certainly never had been. She was a "buxom goblette" and a "sleazy lady" who likes to eat and pop pills.[33] The fan magazine *Blip* described her eyelids as "fluttering," and claimed that when caught by a ghost, Ms. Pac-Man "faints instead of deflating."[34] These are perhaps instances of stretching to find descriptive terms for games that, ten years into the commercial life of the medium, were still fairly abstract in their representation and quite primitive in their character design. But they also look like opportunities to circumscribe female representation by making it different from the norm, and by emphasizing that female identity be evaluated in terms of appearance and sex appeal, even when we're talking about something as simplistic and unsexy as a Pac-Man with lips and a hair bow.

Whatever the impact of this gendered discourse at the time, *Ms. Pac-Man*, like *Pac-Man*, became a favorite game of many girls and women. *Film Comment* mentioned a woman who played *Ms. Pac-Man* exclusively, while a "Talk of the Town" piece in the *New Yorker* described an arcade on Fire Island, the resort near New York City, during the summer of 1982, where girls congregated to play the game. "Ms. Pac-Man players in the Arcade room are mostly gangs of four or five minigirls with one older babysitter— an elder person, usually female."[35] But elsewhere in the same village of Ocean Beach, at Mike the Greek's and an ice cream parlor, *Ms. Pac-Man* was played by adolescent boys "who pound the glass and swear furiously— mostly unmentionables." *Pac-Man* and *Ms. Pac-Man* alike appealed very broadly, and *Ms. Pac-Man* in particular has remained one of the most desirable arcade cabinet games for collectors and public locations for decades, still collecting quarters in bowling alleys, miniature golf courses, and pizza parlors years after its debut.

In the 1980s, the presence of girls and woman among these games' faithful players not only guaranteed their commercial success, but also revealed the assumption that other video games were particularly of interest to young male players—that this was somehow natural rather than a social construct based on the growth of video games within a particular historical context. The novelty of female-favorite games reinforced the common sense that video games were a male preserve. At the moment when video games became omnipresent, mainstream popular media, the gendered reception of *Pac-Man* and *Ms. Pac-Man* showed the newly emerged medium to be safe and friendly and open to all. At the same time, this reception marking the feminine as different also showed that video games were still the way mainly young male players coming of age in the early 1980s worked out their gender identity by continuing in the boy-culture traditions of aggressive, competitive play, updated for the era of *Star Wars* and computers. Having a handful of games with a less intensely masculine character made room for other players without threatening the overwhelming young and male focus on video games more generally.

In the "Pac-Man Fever" episode of the short-lived teen comedy *Square Pegs*, which ran on NBC for one season in 1982–83, the suburban, middle-class, teenage boy Marshall becomes addicted to video games when they appear in the high school cafeteria—he gets addicted to *Pac-Man*—and he loses all interest in the things boys often care about, including girls.[36] Marshall competes against another boy for a high score but neglects his social life and responsibilities because of his video game abuse. His friends recruit an attractive girl to seduce Marshall and break him of his obsession, but to no avail. Not even sexual desire can break the spell. Eventually, after he is comically cured in time for the closing credits, the episode ends on a tag with Marshall's two friends Lauren and Patty encountering a new *Ms. Pac-Man* cabinet at the diner ("shouldn't that be *Ms. Pac-Person?!*"). One asks the other for a quarter as she approaches the machine. This suggests a full-circle, "Here we go again" ending, but this time with a teenage girl having a turn at getting lost in this new form of seductive play. Of course, the focus of the whole episode on a boy getting hooked underscores the normative gendering of video games, while the humorous twist at the end gestures at inclusiveness. But surely *Ms. Pac-Man* was an opportunity for many girls in the 1980s to experience video games on their own terms rather than as a foray into boy culture. *Pac-Man* and *Ms. Pac-Man* allowed for this expansion of video games to make room for female players, but it did so without seriously challenging the deeper inequality within the new culture of electronic leisure.

Conclusion

Pac-Man Fever marks a culmination. By the time of *Square Pegs*, it's fair to say, the medium of video games had closed off much of the flexibility of its meanings. What had been a newfangled gadget or novelty was by the fall of 1982 a fixture of everyday life, an unavoidable presence permeating pop culture. Pac-Man Fever could be caught by anyone. It was truly mass media, but this was in tension with a cloistered quality in the world of video games, a boy's-club ideal that has endured decades after stores stopped selling Pac-Man socks and sweatshirts. Tensions between inclusiveness and exclusiveness, a core of subcultural fan identity and mainstream popularity, masculine fandom and broader appeals, have been an enduring legacy of early video games.

Like any emergent medium or technology, video games were not invented out of nothing, and they were never merely a neutral artifact to be used in whatever ways individuals desired. They came along within a set of long-standing social relations characterized by differences in power, authority, and opportunity. These differences structured social identities such as age, race, class, gender, and sexuality, and also dynamics in public places and in the home. Ideas about advanced technology, coin-operated amusements, and domestic recreation from long before the invention and emergence of electronic games shaped the development of the medium. Ideas about middle-class values, about forms of play for boys and girls, and about coming of age carried over from earlier times and earlier experiences of leisure. Contours of suburban and urban spaces, Cold War politics, transformations in labor and economics, the computerization of society, and fashions and styles in popular culture all left imprints on video games. Games were always understood through points of reference to already familiar objects and experiences, and to new ones emerging alongside high-tech play: spaceships real and imagined, TV sets and their possible uses, recreation rooms and the activities within them, arcades dusty or clean, computers old and new, and other toys with microchips inside. All of these objects and experiences came burdened with associations and expectations.

As the response to *Pac-Man's* wild popularity shows, video games emerged into a defined cultural profile. Of course, not all players of video games in these years were young (teenage and young adult), male, and middle-class, but such players formed the core ideal of the medium in relation to its users. The material culture of early video games, the games as objects, their images, sounds, and gameplay, and representations of them in many media, contributed to this ideal, this widely shared sense of the medium's value and its typical uses and pleasures. So did ideas imposed by guardians of morality and businesses standing to profit from games and their players. Both the games and the ways they were experienced contributed to a widely shared sense of their cultural status, a place in popular imagination. It was not inevitable that games should take on the identity that they did. A myriad of individual intentions, social forces, and background assumptions shaped this development. Later efforts to expand and renegotiate our expectations about video games and what they can and should be—and for whom—still struggle against the meanings worked out during this crucial formative period of emergence.

Notes

Preface

1. Michael Z. Newman and Elana Levine, *Legitimating Television: Media Convergence and Cultural Status* (New York: Routledge, 2011); Michael Z. Newman, "Free TV: File-Sharing and the Value of Television," *Television and New Media* 13, no. 6 (2012): 463–479.

2. Work published several years after I began my research includes Carly Kocurek, *Coin-Operated Americans: Rebooting Boyhood at the Video Arcade* (Minneapolis: University of Minnesota Press, 2015).

3. Tom Gunning, "The Cinema of Attraction: Early Film, Its Spectator, and the Avant-Garde," *Wide Angle* 3, no. 4 (1986): 63–70; Charles Musser, *The Emergence of Cinema: The American Screen to 1907* (Berkeley: University of California Press, 1994); Thomas Elsaesser, ed., *Early Cinema: Space Frame Narrative* (London: BFI, 1990); William Boddy, *Fifties Television: The Industry and Its Critics* (Champaign: University of Illinois Press, 1990); Lynn Spigel, *Make Room for TV: Television and the Family Ideal* (Chicago: University of Chicago Press, 1992); Cecilia Tichi, *Electronic Hearth: Creating an American Television Culture* (New York: Oxford University Press, 1992).

Introduction: Early Video Games and New Media History

1. Some key scholarly contributions to new media studies are Jay David Bolter and Richard Grusin, *Remediation: Understanding New Media* (Cambridge, MA: MIT Press, 1998); Lisa Gitelman, *Always Already New: Media, History, and the Data of Culture* (Cambridge, MA: MIT Press, 2006); Henry Jenkins, *Convergence Culture: Where Old and New Media Collide* (New York: NYU Press, 2006); and Lev Manovich, *The Language of New Media* (Cambridge, MA: MIT Press, 2001). For one challenge to the widespread use of the term, see Tim Anderson, "'New Media'? Please Define," *Flow*, May 12, 2006, http://www.flowjournal.org/2006/05/new-media-please-define/ (accessed April 2, 2016).

2. William Boddy, *New Media and Popular Imagination: Launching Radio, Television, and Digital Media in the United States* (New York: Oxford University Press, 2004); Benjamin Peters, "Lead Us Not into Thinking the New Is New: A Bibliographic Case for New Media History," *New Media and Society* 11, nos. 1–2 (2009): 13–30.

3. Trevor J. Pinch and Wiebe E. Bijker, "The Social Construction of Facts and Artefacts: Or How the Sociology of Science and the Sociology of Technology Might Benefit Each Other," *Social Studies of Science* 14 (1984): 399–441; Ronald Kline and Trevor Pinch, "Users as Agents of Technological Change: The Social Construction of the Automobile in the Rural United States," *Technology and Culture* 37, no. 4 (1996): 763–795.

4. Bolter and Grusin, *Remediation*.

5. On gamer identity see Adrienne Shaw, "Do You Identify as a Gamer? Gender, Race, Sexuality, and Gamer Identity," *New Media and Society* 14 (2012): 28–44; Adrienne Shaw, "On Not Becoming Gamers: Moving beyond the Constructed Audience," *Ada: A Journal of Gender, New Media, and Technology* 2 (2013), doi: 10.7264/N33N21B3 (accessed April 5, 2016).

6. Roberto Dillon, *The Golden Age of Video Games: The Birth of a Multi-Billion Dollar Industry* (Boca Raton: CRC Press, 2011); Tristan Donovan, *Replay: The History of Video Games* (East Sussex, UK: Yellow Ant, 2010); Steven L. Kent, *The Ultimate History of Video Games* (Roseville, CA: Prima Publishing, 2001).

7. Carly Kocurek, *Coin-Operated Americans: Rebooting Boyhood at the Video Game Arcade* (Minneapolis: University of Minnesota Press, 2015), was published after I had completed a draft of this book. Kocurek offers engaging analytical arguments about gender and video games in the context of American society shifting from industrial to postindustrial labor and production.

8. On *Tennis for Two* and the Brown Box, see Raiford Guins, *Game After: A Cultural Study of Video Game Afterlife* (Cambridge, MA: MIT Press, 2014). I discuss *Spacewar!* in chapter 4.

9. Leonard Herman, "Ball-and-Paddle Consoles," in *Before the Crash: Early Video Game History*, ed. Mark J. P. Wolf (Detroit: Wayne State University Press, 2012), 53–59.

10. "New Tricks Your TV Can Do," *Changing Times*, October 1976, 19–20.

11. Henry Lowood, "Videogames in Computer Space: The Complex History of Pong," *IEEE Annals of the History of Computing* 31, no. 1 (2009): 5–19.

12. Bernice Kanner, "Can Atari Stay Ahead of the Game?" *New York*, August 16, 1982, 15–17; Peter Nulty, "Why the Craze Won't Quit," *Fortune*, November 15, 1982, 114–124.

13. Donovan, *Replay*, 97–108; Gary Putka, "Warner Communications Sends Host of Stocks Tumbling with Its Reduced Earnings Estimate," *Wall Street Journal*, December 9, 1982, 55; "The Real Trouble in Video Games," *Business Week*, December 27, 1982, 31–32.

14. Michael Z. Newman, "When Television Marries Computer," *Flow*, November 18, 2013, http://www.flowjournal.org/2013/11/when-television-marries-computer/.

15. Nancy Baym, *Personal Connections in the Digital Age* (London: Polity, 2010); Daniel J. Czitrom, *Media and the American Mind: From Morse to McLuhan* (Chapel Hill: University of North Carolina Press, 1982), Claude S. Fischer, *America Calling: A Social History of the Telephone to 1940* (Berkeley: University of California Press, 1992), Michèle Martin, "The Culture of the Telephone," in *Sex/Machine: Readings in Culture, Gender and Technology*, ed. Patrick D. Hopkins (Bloomington: Indiana University Press, 1989), 50–74; Carolyn Marvin, *When Old Technologies Were New: Thinking about Electric Communication in the Late Nineteenth Century* (New York: Oxford University Press, 1988); Spigel, *Make Room for TV*.

16. Kristen Drotner, "Dangerous Media? Panic Discourses and Dilemmas of Modernity," *Paedagogica Historica* 35, no. 3 (1999): 593–619.

17. Newman, *Video Revolutions*.

18. Geoffrey R. Loftus and Elizabeth F. Loftus, *Mind at Play: The Psychology of Video Games* (New York: Basic Books, 1983); David Sudnow, *Pilgrim in the Microworld* (New York: Warner Books, 1983); Sherry Turkle, *The Second Self: Computers and the Human Spirit* (New York: Simon and Schuster, 1984).

19. The idea of meaning being made through the reader's horizon of expectations is from Hans Robert Jauss and Elizabeth Benzinger, "Literary History as a Challenge to Literary Theory," *New Literary History* 2, no. 1 (1970): 7–37.

20. Lynn Spigel, *Welcome to the Dreamhouse: Popular Media and Postwar Suburbs* (Durham, NC: Duke University Press, 2001), 15.

1   Good Clean Fun: The Origins of the Video Arcade

1. Carly Kocurek, "Coin-Drop Capitalism: Economic Lessons from the Video Arcade," in *Before the Crash: Early Video Game History*, ed. Mark J. P. Wolf (Detroit: Wayne State University Press, 2012), 189–208; Ryan Pierson, "Making Sense of Early Video Arcades: The Case of Pittsburgh, 1980–1983," *Canadian Journal of Film Studies* 20, no. 2 (2011): 19–37.

2. For an illustrated celebration of these games and the spaces in which they were played, see Van Burnham, *Supercade: A Visual History of the Videogame Age, 1971–1983* (Cambridge, MA: MIT Press, 2001). An essay drawing revealing connections between generations of arcade spaces is Erkii Huhtamo, "Slots of Fun, Slots of

Trouble: An Archaeology of Arcade Gaming," in *Handbook of Computer Game Studies*, ed. Joost Raessens and Jeffrey Goldstein (Cambridge, MA: MIT Press, 2005), 3–21.

3. For a history of media reports framing video games in their early years, see Dmitri Williams, "The Video Game Lightning Rod: Constructions of a New Media Technology, 1970–2000," *Information, Communication, and Society* 6, no. 4 (2003): 523–550. I return to the perceived threats posed by arcades in further detail in chapter 5.

4. For instance, the video-game-themed February 26, 1983, episode of ABC's public affairs program *Nightline* raised concerns about children spending too much money playing coin-operated games, about the games causing violence, about their addictive nature, and about the player's absorption in the game. For more, see chapter 5.

5. Paul Blustein, "Video-Game Makers Zap Surgeon General Over Health Remark," *Wall Street Journal*, November 11, 1982.

6. "The Slot," *Illustrated American*, January 3, 1891, 222–223.

7. Rollin Lynde Hartt, *The People at Play* (Boston: Houghton Mifflin, 1909); Roger C. Sharpe, *Pinball!* (New York: E. P. Dutton, 1977).

8. Brooks McNamara, "'A Congress of Wonders': The Rise and Fall of Dime Museums," *ESQ: A Journal of the American Renaissance* 20, no. 3 (1974): 216–232; Andrea Stulman Dennett, *Weird and Wonderful: The Dime Museum in America* (New York: NYU Press, 1997).

9. "The Penny Arcade," *Chicago Tribune*, August 22, 1903, 6.

10. Michael M. Davis, Jr., *The Exploitation of Pleasure: A Study of Commercial Recreations in New York City* (New York, 1911).

11. Hartt, *The People at Play*.

12. Charles Musser, *The Emergence of Cinema: The American Screen to 1907* (Berkeley: University of California Press, 1990), 176.

13. Davis, *The Exploitation of Pleasure*, 10.

14. "The Penny Arcade."

15. Hartt, *The People at Play*, 120.

16. David Nasaw, *Going Out: The Rise and Fall of Public Amusements* (New York: Basic Books, 1993), 120–134.

17. Bertram Reinitz, "Penny Pleasures Still," *New York Times*, August 9, 1925.

18. Davis, *The Exploitation of Pleasure*.

19. "'Penny Arcade' Exhibit Due," *New York Times*, October 8, 1961.

20. Davis, *The Exploitation of Pleasure*.

21. Sharpe, *Pinball!*, 48.

22. Bob Levitt, "Rise of Game Emporiums," *Billboard*, October 14, 1933.

23. Sharpe, *Pinball!*, 35.

24. Edward Trapunski, *Special When Lit: A Visual and Anecdotal History of Pinball* (Garden City, NY: Dolphin Books, 1979), 30.

25. *United States v. Korpan*, 237 F. 2d 676—Court of Appeals, 7th Circuit, 1956.

26. "Pinball—Depression Smasher!" *Billboard*, January 18, 1936.

27. Quoted in Trapunski, *Special When Lit*, 98.

28. Ibid., 96.

29. The *All in the Family* episode "Archie's Grand Opening," season 8, episode 7, originally aired November 6, 1977.

30. Stephen Peeples, "A Word from the Players," *RePlay*, May 1977, 36–37.

31. John McPhee, "The Pinball Philosophy," *New Yorker*, June 30, 1975, 81–83.

32. These include Sharpe, *Pinball!* and Trapunski, *Special When Lit*.

33. Trapunski, *Special When Lit*, 38.

34. David A. Cook, *Lost Illusions: American Cinema in the Shadow of Vietnam and Watergate, 1970–1979* (New York: Scribner's, 2000), 307–309.

35. Trapunski, *Special When Lit*, 117.

36. Charles Walker, "Amusement Centers: Still Problems but Their Early Image Is Changing," *Shopping Center World*, April 1978, 22–25.

37. On malls as spaces for suburban teenagers to spend their leisure time and claim as their own during the 1980s in particular, see Kyle Riismandel, "Arcade Addicts and Mallrats: Producing and Policing Suburban Public Space in 1980s America," *Environment, Space, Place* 5, no. 2 (2013): 65–89.

38. William Severini Kowinski, *The Malling of America: An Inside Look at the Great Consumer Paradise* (New York: William Morrow, 1985), 36.

39. Sidney J. Kaplan, "The Image of Amusement Arcades and Differences in Male and Female Video Game Playing," paper presented at the 1982 annual meeting of the North Central Sociological Association.

40. "1976 *RePlay* Arcade Analysis," *RePlay*, March 1976, 21–24.

41. Roger Sharpe, "U.S. Arcades: Some Good, Some Bad, Some Ugly," *RePlay*, March 1977, 39–40.

42. "The Year That Was 1980: A Nation of Inflation and Game Proliferation," *RePlay*, January 1981, 25–34.

43. "Murph Gordon of Aladdin's Castle: Spreading the Arcade from Coast to Coast," *RePlay*, November 1979, 43–46.

44. Victor Gruen and Larry Smith, *Shopping Towns USA: The Planning of Shopping Centers* (New York: Reinhold, 1960), 23.

45. James W. Rouse, "Must Shopping Centers Be Inhuman?" *Architectural Forum*, June 1962, 105–107, 196; followed by Rouse, "Center for Rochester," 108–112. See also Gruen and Smith, *Shopping Towns USA*, 257–258.

46. Peter O. Muller, Contemporary Suburban America (Englewood Cliffs, NJ: Prentice-Hall, 1981), 125–126; Neil Harris, "Spaced Out at the Shopping Center," *Cultural Excursions: Marketing Appetites and Cultural Tastes in Modern America* (Chicago: University of Chicago Press, 1990), 278–288.

47. Jon Goss, "The 'Magic of the Mall': An Analysis of Form, Function, and Meaning in the Contemporary Retail Built Environment," *Annals of the Association of American Geographers* 83, no. 1 (1993), 18–47; Kowinski, *The Malling of America*, 58–62.

48. Goss, "The 'Magic of the Mall,'" 26.

49. On the history of suburbanization and its racial and class dynamics, see Kenneth Jackson, *Crabgrass Frontier: The Suburbanization of the United States* (New York: Oxford University Press, 1987).

50. "Editorial: Outlook '77," *RePlay*, January 1977, 3; "GNS Opens Biggie in Phoenix," *RePlay*, December 1977, 29; "The Seventies: A Dynamic Decade in Industry History," *RePlay*, December 1979, 7–24.

51. Walker, "Amusement Centers"; Trapunski, *Special When Lit*, 147; Sharp, *Pinball!*, 85; Al Rodstein, "Arcades—Much More Room to Grow," *RePlay*, December 20, 1975, 10; Louis Boasberg, "The Rise and Fall (and Rise) of the Arcade," *RePlay*, December 20, 1975, 10–12; Gene Castellano, "'Tilt' the Movie: Maybe Not as Bad as We Expected?" *RePlay*, March 1979, 51–54; "Good Press Fuels Consumer Love of Games," *RePlay*, September 1979, 38–43.

52. "Editorial: Powderkeg," *RePlay*, April 1977, 3.

53. Karl Kolad, "Glean Fortune in Pennies That Are Dropped in Slots," *Chicago Tribune*, January 28, 1906, E1.

54. Bob Levitt, "Rise of Game Emporiums," *Billboard*, October 14, 1933.

55. "Editorial: Powderkeg"; Rodstein, "Arcades"; Boasberg, "The Rise and Fall."

56. "RePlay Reports: The 1976 Equipment vs. Collection Review," *RePlay*, September 1976, 6–7; "1978 Route & Arcade Survey," *RePlay*, November 1978, 41–48;

"Editorial," *RePlay*, April 1982, 3; Carol Kantor, "Good Guys Sometimes Get the Blame," *RePlay*, May 1982, 38; "Boasberg's Rules for Good Game Centers: Self-Policing Guidelines Still a Must," *RePlay*, May 1982, 41–42.

57. "Plush Arcades Attract Families to Time Zone," *RePlay*, December 1976, 26–27.

58. "Editorial: Powderkeg."

59. "1978 Route & Arcade Survey."

60. "Plush Arcades Attract Families to Time Zone"; "Editorial: Outlook '77"; Marshall Caras, "'78 Speculation: The Year of the Great Shake Out," *RePlay*, January 1978, 33–34; "'78 Collection Report," *RePlay*, September 1978, 12; "Castle Park Arcade," *RePlay*, March 1981, 39–40.

61. "Arcades Today," *RePlay*, September 1978, 26.

62. "Editorial: Outlook '77."

63. "Arcade Inn: Arcades Flourish in Ramada Inn Chain Thanks to Jeff Lipsman's Watchful Eye," *RePlay*, April 1979, 61.

64. "1977 Route & Arcade Survey," *RePlay*, October 1977, 19–26; Marshall Caras, "And Then There Were Three," *RePlay*, December 1977, 11–12; "The RePlay Operator Survey," *RePlay*, November 1979, 90–100.

65. Boasberg, "The Rise and Fall."

66. "1977 Route & Arcade Survey."

67. "Street and Arcade Operator Survey," *RePlay*, March 1977, 26–28.

68. "1976 RePlay Arcade Analysis," *RePlay*, March 1976, 21–24.

69. "Street and Arcade Operator Survey."

70. "On the Line," *RePlay*, August 1978, 14.

71. Louis Boasberg, "TV Video Games and Respectability," *RePlay*, October 30, 1976, 20–21.

72. Ibid.

73. "Such Love! The Best PR of All," *RePlay*, January 4, 1978, 40.

74. "Plush Arcades Attract Families to Time Zone."

75. "Westworld Arcade Gala Opening," *RePlay*, October 1977, 22.

76. "'Tilt' the Movie."

77. "Marketing the World's Hottest Game," *RePlay*, April 1979, 17.

78. Marshall Caras, "Video and the Jukebox: The Old Rules May Be Shattered," *RePlay*, January 1981, 47.

79. "The Year That Was 1980."

80. "Operating Videos Is Serious Hardball," *RePlay*, July 1981, 21–24.

81. "Equipment Poll," *Play Meter*, March 1, 1982, 10.

82. "The Year That Was 1980."

83. "Atari's Mighty 'Asteroids' Space Target Video Hits," *RePlay*, February 1980, 35.

84. Carly Kocurek, *Coin-Operated Americans: Rebooting Boyhood at the Video Game Arcade* (Minneapolis: Minnesota University Press, 2015), 44–47.

85. Valerie Cognevich, "They Play for Hours, and Hours, and Hours to Beat the High Score!" *Play Meter*, December 1, 1981, 163–164.

86. "Youngsters Tell RePlay the Role the Games Play in Their Lives," *RePlay*, August 1981, 34–51.

87. Sherry Turkle, *The Second Self: Computers and the Human Spirit* (New York: Simon & Schuster, 1984), 64–92.

2 "Don't Watch TV Tonight. Play It!" Early Video Games and Television

1. "Instant Replay," *Newsweek*, October 30, 1972, 75.

2. Michael Z. Newman, *Video Revolutions: On the History of a Medium* (New York: Columbia University Press, 2014). In this work, I trace this history and transformation of "video" from being another word for TV to becoming a term more often used to refer to something distinct from television as a commercial broadcasting medium.

3. Robert Wieder, "A Fistful of Quarters," *Oui*, 1974, 59–62, 124–129; 128.

4. Raymond Williams, *Television: Technology and Cultural Form* (Hanover, NH: Wesleyan University Press, 1974).

5. Daniel Bell, "The Theory of Mass Society: A Critique" (1956), in *Mass Communication and American Social Thought: Key Texts 1919–1968*, ed. John Durham Peters and Peter Simonson (Lanham: Rowman & Littlefield, 2014), 364–373; C. Wright Mills, "The Mass Society," in Durham Peters and Simonson (eds.), *Mass Communication*, 387–400; Bernard Rosenberg and David Manning White, eds., *Mass Culture: The Popular Arts in America* (Glencoe, IL: The Free Press, 1958); Edward Shils, "Mass Society and Its Culture," *Deadalus* 89, no. 2 (1960): 288–314.

6. Two recent studies of twentieth-century media that engage insightfully with this history are Charles R. Acland, *Swift Viewing: The Popular Life of Subliminal Influence* (Durham, NC: Duke University Press, 2012), and Fred Turner, *The Democratic*

Surround: Multimedia and American Liberalism from World War II to the Psychedelic Sixties (Chicago: University of Chicago Press, 2013).

7. Theodor W. Adorno and Max Horkheimer, "The Culture Industry: Enlightenment as Mass Deception," in *Dialectic of Enlightenment: Philosophical Fragments*, trans. Edmund Jephcott (Palo Alto, CA: Stanford University Press, 1944/2002).

8. W. Russell Neuman, *The Future of the Mass Audience* (Cambridge: Cambridge University Press, 1991), 23–26.

9. C. Wright Mills, *The Power Elite* (New York: Oxford University Press, 1956). Page numbers to follow are from the chapter "The Mass Society" anthologized in Durham Peters and Simonson (eds.), *Mass Communication*, 387–400.

10. Ibid., 389.

11. Ibid., 390.

12. Ibid., 391.

13. Ibid., 395–396.

14. C. Wright Mills, "Letter to the New Left," *New Left Review* 5 (1960).

15. "Port Huron Statement of the Students for a Democratic Society, 1962," http://www.h-net.org/~hst306/documents/huron.html, accessed August 11, 2015.

16. Milton Viorst, *Fire in the Streets: America in the 1960s* (New York: Simon & Schuster, 1979), 191. See also Andrew Hunt, "How New Was the New Left? Re-Thinking New Left Exceptionalism," in *The New Left Revisited*, ed. John McMillian and Paul Buhle (Philadelphia: Temple University Press, 2003), 139–155.

17. George Movshon, "The Video Revolution," *Saturday Review*, August 8, 1970, 50–52; Brenda Maddox, *Beyond Babel: New Directions in Communication* (New York: Simon & Schuster, 1972), 146–147.

18. Jay David Bolter and Richard Grusin, *Remediation: Understanding New Media* (Cambridge, MA: MIT Press, 2000).

19. Henry Lowood, "Video Games in Computer Space: A Complex History of *Pong*," *IEEE Annals of the History of Computing*, July–September 2009, 5–19; 6.

20. "Magnavox Unveils TV Game Simulator," *New York Times*, May 11, 1972.

21. Cindy Morgan, "Video Games: Put Your Backhand on TV," *Popular Mechanics*, October 1972, 78–80.

22. Larry Steckler, "TV Games at Home," *Radio-Electronics*, December 1975, 29.

23. "Odyssey, a Game You Play through Your TV Set," *Consumer Reports*, February 1973, 81–3.

24. "TV's Hot New Star: The Electronic Game," *Business Week*, December 29, 1975, 24.

25. Sheila C. Murphy, *How Television Invented New Media* (New Brunswick, NJ: Rutgers University Press, 2011), 46.

26. Ralph H. Baer, *Videogames in the Beginning* (Springfield, NJ: Rolenta Press, 2005), 19.

27. Lynn Spigel, *Make Room for TV: Television and the Family Ideal in Postwar America* (Chicago: University of Chicago Press, 1992).

28. Murphy, *How Television Invented New Media*, 50.

29. "How the Two Big TV Toys Are Faring," *Broadcasting*, January, 24, 1977, 80.

30. Ibid.

31. Movshon, "The Video Revolution."

32. Roger Kenneth Field, "In the Sixties, It Was TV; In the Seventies, Video Cassette," *New York Times*, June 6, 1970.

33. Max Dawson, "Home Video and the 'TV Problem': Cultural Critics and Technological Change," *Technology and Culture* 48 (2007): 524–549; 526.

34. Robert Chew, "Innovations in Video—Nightmare for Networks?" *Advertising Age*, May 30, 1977, 3, 70.

35. "Instant Replay."

36. "New Tricks Your TV Can Do," *Changing Times*, October 1976, 10, 30.

37. For instance, see Josh Levine, "Mattel Eyes Software, Cable TV Linkup," *Advertising Age*, February 26, 1979, 3, 81.

38. Maddox, *Beyond Babel*, 145–146.

39. Ralph Lee Smith, *The Wired Nation* (New York: Harper Colophon, 1972), 7–8.

40. Thomas Streeter, "Blue Skies and Strange Bedfellows: The Discourse of Cable Television," in *The Revolution Wasn't Televised: Sixties Television and Social Conflict*, ed. Lynn Spigel and Michael Curtin (New York: Routledge, 1997), 221–242; 236–237.

41. Murphy, *How Television Invented New Media*, 50.

42. Howard Gardner, "When Television Marries Computer," *New York Times*, March 27, 1983.

43. Michael Z. Newman and Elana Levine, *Legitimating Television: Media Convergence and Cultural Status* (New York: Routledge, 2012), 5.

44. Tristan Donovan, *Replay: The History of Video Games* (East Sussex, UK: Yellow Ant Media, 2010), 8.

45. Quoted in Jason Wilson, "'Participation TV': Videogame Archaeology and New Media Art," in *The Pleasures of Computer Gaming: Essays on Cultural History, Theory, and Aesthetics*, ed. Melanie Swalwell and Jason Wilson (Jefferson, NC: MacFarland, 2008), 94–117; 106.

46. Judy Klemesrud, "Bang! Boing! Ping! It's King Pong," *New York Times*, April 24, 1978.

47. Donovan, *Replay*, 36.

48. "The Big Winners in Consumer Sales," *Sales & Marketing Management*, August 9, 1976, 23–25.

49. "Magnavox Unveils TV Game Simulator."

50. Peter Ross Range, "The Space-Age Pinball Machine," *New York Times Magazine*, September 15, 1974.

51. "Modern Living: Screen Games," *Time*, May 22, 1972.

52. Larry Steckler, "TV Games at Home," *Radio Electronics*, December 1975, 29–31, 71, 90–91; 29.

53. Dick Pietschmann, "The New Fun World of Video Games," *Mechanix Illustrated*, January 1975, 36, 92; 36.

54. William D. Smith, "Electronic Games Bringing a Different Way to Relax," *New York Times*, December 25, 1975.

55. Carll Tucker, "Sociable Pong," *Saturday Review*, November 26, 1977, 56.

56. David Sudnow, *Pilgrim in the Microworld* (New York: Warner Books, 1983), 69.

57. Wilson, "Participation TV," 97–100.

58. Ibid., 98.

59. Marita Sturken, "TV as a Creative Medium: Howard Wise and Video Art," *AfterImage*, May 1984, 5–9.

60. Wilson, "Participation TV," 112.

61. "Odyssey—It Turns Your TV into an Action Game," *Popular Science*, August 1972, 54; "Business Journal," *Wall Street Journal*, December 9, 1976.

62. John Crudele, "Consumer Electronics Home $weet Home," *Electronic News*, April 4, 1977, 22–26; 22.

3   Space Invaders: Masculine Play in the Media Room

1. Bernice Kammer, "Can Atari Stay Ahead of the Game?" *New York*, August 16, 1982, 15–17.

2. "Pinball Redux: The Hottest Games," *Time*, October 31, 1977.

3. Nancy K. Baym, *Personal Connections in the Digital Age* (Malden, MA: Polity, 2010), 45–48; Thomas Berker, Maren Hartmann, Yves Punie, and Katie Ward., eds., *Domestication of Media and Technology* (Maidenhead, UK: Open University Press, 2006).

4. Lynn Spigel, *Welcome to the Dreamhouse: Popular Media and Postwar Suburbs* (Durham, NC: Duke University Press, 2001), 11.

5. Erkki Huhtamo, "What's Victoria Got to Do with It? Toward an Archaeology of Domestic Video Gaming," in *Before the Crash: Early Video Game History*, ed. Mark J. P. Wolf (Detroit: Wayne State University Press, 2012), 30–52; 48.

6. Witold Rybczynski, *Home: A Short History of an Idea* (New York: Viking Penguin, 2001).

7. Thomas J. Schlereth, *Victorian America: Transformations in Everyday Life, 1876–1915* (New York: HarperCollins, 1991), 122–123.

8. Cle Kinney and Barry Roberts, *Don't Move—Improve! Hundreds of Ways to Make a Good House Better* (New York: Thomas Y. Crowell, 1979), 89.

9. Lynn Spigel, *Make Room for TV: Television and the Family Ideal in Postwar America* (Chicago: University of Chicago Press, 1992).

10. George Nelson, *Tomorrow's House: A Complete Guide for the Home-Builder* (New York: Simon & Schuster, 1945), 76–80.

11. Ibid., 77.

12. Ibid., 80.

13. "Playrooms in Suburban Residences Strike a Note of Informality in Different Ways," *New York Times*, July 15, 1956, 1.

14. Clifford J. Clark, Jr., *American Family Home, 1800–1960* (Chapel Hill: University of North Carolina Press, 1986), 203.

15. Lisa Jacobson, *Raising Consumers: Children and the American Mass Market in the Early Twentieth Century* (New York: Columbia University Press, 2005).

16. Cited in Jacobson, *Raising Consumers*, 161.

17. Ibid., 176.

18. Ibid., 180. Jacobson cites Donna R. Braden, "'The Family That Plays Together Stays Together': Family Pastimes and Indoor Amusements, 1890–1930," in *American Home Life*, 148.

19. Ibid., 181.

20. Wayne Whittaker, "Introducing the *PM* Young-Family House," *Popular Mechanics*, October 1955, 153–156.

21. "Builders Study Playroom's Role as a Guide to Planning a Home," *New York Times*, August 25, 1957, 1.

22. "Playrooms in Suburban Residences Strike a Note of Informality in Different Ways," *New York Times*, July 15, 1956, 1.

23. Annette Sukov, "Basement Rec Rooms with a Family Room Look," *Popular Mechanics*, April 1973, 141–143.

24. Richard Stepler, "For a Dramatic Entry to Your Family Room, Dig Yourself a Stairwell," *Popular Science*, June 1975, 94–95.

25. "Do Home Improvements Pay Off?" *Changing Times*, December 1974, 13–15.

26. Shirley Maxwell and James C. Massey, "The Modern Basement: A Matter of Ups and Downs," *Old House Journal*, May–June 1992, 30–33.

27. Ralph Treves, *How to Make Your Own Recreation and Hobby Rooms*, 2nd ed. (New York: Harper & Row, 1976).

28. Kinney and Roberts, *Don't Move—Improve!*, 97.

29. Dick Pietschmann, "The New Fun World of Video Games," *Mechanix Illustrated*, January 1975, 36; Jim Stickford, "How to Do Your Basement in Pub Decor," *Mechanix Illustrated*, January 1975, 37.

30. Keitha McLean, "The Good Life: Media Room," *American Home*, November 1976, 38–39.

31. Joan Kron, "The Media Room," *New York*, April 19, 1976, 55–61.

32. Patricia Delaney, "Entertainment on the House," *Time*, July 27, 1981.

33. "Media Room: Entertainment Center in Your Home," *Popular Mechanics*, December 1982, 86–88, 111.

34. Delaney, "Entertainment on the House."

35. "Media Room: Entertainment Center in Your Home."

36. George O'Brien, "Living with Electronics," *New York Times Magazine*, September 27, 1981, 25–28.

37. Maury Z. Levy, "The Video Explosion," *Cincinnati*, October 1977.

38. O'Brien, "Living with Electronics."

39. Michael George, "Magnificent Media Rooms," *Black Enterprise*, January 1984, 58–60.

40. "Odyssey ... It Turns Your TV into an Action Game," *Popular Science*, August 1972, 54.

41. Dick Pietschmann, "The Fun New World of Video Games," *Mechanix Illustrated*, January 1975, 36, 92.

42. Larry Steckler, "TV Games at Home," *Radio Electronics*, December 1975, 29–31, 71, 90–91.

43. Kris Jensen, "New 1978 Electronic Games," *Popular Electronics*, January 1978, 33–41.

44. "Home Video Warfare Erupts on Television," *Broadcasting*, March 1, 1982, 64.

45. My discussion of boy culture follows the usage of E. Anthony Rotundo, "Boy Culture," in *The Children's Culture Reader*, ed. Henry Jenkins (New York: NYU Press, 1998), 337–362.

46. Edna Mitchell, "The Dynamics of Family Interaction around Home Video Games," *Marriage and Family Review* (1985): 121–135; 123.

47. Kay Rohl Murphy, "Family Patterns of Use and Parental Attitudes toward Home Electronic Video Games and Future Technology," PhD diss., Oklahoma State University, 1984, 105–106.

48. Mitchell, "The Dynamics of Family Interaction," 129.

49. Ibid., 130.

50. Brian Sutton-Smith, *Toys as Culture* (Dalton, OH: Gardner Press, 1986), 66. The conference's proceedings were published as *Video Games and Human Development: A Research Agenda for the 80s* (Cambridge, MA: Monroe C. Gutman Library, 1983). I discuss this conference and the research presented there in more detail in chapter 5.

51. Ralph J. Watkins, "A Competitive Assessment of the U.S. Video Game Industry," United States International Trade Commission, Washington, DC, March 1984.

52. Ibid., 42.

53. Murphy, "Family Patterns of Use," 124.

54. Watkins, "A Competitive Assessment," 44.

55. Sutton-Smith, *Toys as Culture*, 61.

56. Ibid., 69.

57. Steven Mintz, *Huck's Raft: A History of American Childhood* (Cambridge, MA: Belknap Press of Harvard University Press, 2006), 83.

58. Ibid.

59. Rotundo, "Boy Culture."

60. Mintz, *Huck's Raft*, 218.

61. The images and sound files for *The Story of Atari Breakout* can be found at http://www.codehappy.net/breakout/ (accessed July 6, 2016).

62. Mintz, *Huck's Raft*, 347.

63. Marta Gutman and Ning de Coninck-Smith, *Designing Modern Childhoods: History, Space and the Material Culture of Children* (New Brunswick, NJ: Rutgers University Press, 2008), 5; Alison J. Clark, "Coming of Age in Suburbia: Gifting the Consumer Child," in *Designing Modern Childhoods*, 253–268.

64. Henry Jenkins, "Complete Freedom of Movement: Video Games as Gendered Play Spaces," in *The Game Design Reader: A Rules of Play Anthology*, ed. Katie Salen and Eric Zimmerman (Cambridge, MA: MIT Press, 2006), 330–363.

65. Bernadette Flynn, "Geographies of the Digital Hearth," *Information, Communication & Society* 6, no. 4 (2003): 551–576; 558.

4   Video Games as Computers, Computers as Toys

1. William D. Marbach with Peter McAlevey, "A New Galaxy of Video Games," *Newsweek*, October 25, 1982, 123–125; 125.

2. Les Spindle, "Games (Computer) People Play," *Radio-Electronics*, March 1982, 76–77.

3. Tristan Donovan, *Replay: The History of Video Games* (East Sussex, UK: Yellow Ant, 2010), 11.

4. On the development of the personal computer, see Paul E. Ceruzzi, *A History of Modern Computing*, 2nd ed. (Cambridge, MA: MIT Press, 2003), 207–242; Paul Freiberger and Michael Swaine, *Fire in the Valley: The Making of the Personal Computer* (Berkeley, CA: Osborne/McGraw-Hill, 1984); Ted Friedman, *Electric Dreams: Computers in American Culture* (New York: NYU Press, 2005), 81–120; Leslie Haddon, "The Home Computer: The Making of a Consumer Electronic," *Science as Culture* 2 (1988): 7–51; Steven Levy, *Hackers: Heroes of the Computer Revolution* (New York: Anchor, 1984); and John Markoff, *What the Dormouse Said: How the Sixties Counterculture Shaped the Personal Computer Industry* (New York: Viking, 2005).

5. Martin Campbell-Kelly, *From Airline Reservations to Sonic the Hedgehog: A History of the Software Industry* (Cambridge, MA: MIT Press, 2003), 212–214; Christopher Felix

McDonald, "Building the Information Society: A History of Computing as a Mass Medium," PhD diss., Princeton University, 2011.

6. Ted Nelson, *Computer Lib/Dream Machines*, self-published, 1974; Fred Turner, *From Counterculture to Cyberculture: Stewart Brand, the Whole Earth Network, and the Rise of Digital Utopianism* (Chicago: University of Chicago Press, 2006).

7. Stephen H. Seidman, "Hobby and Game Markets Fade … Is There Life after Kits for Retailers?" *Datamation*, April 1979, 80–82.

8. Throughout I am using the term "flexibility," as well sometimes as the phrase "interpretive flexibility," following the Social Construction of Technology approach to the history of technological artifacts. See Trevor J. Pinch and Wiebe E. Bijker, "The Social Construction of Facts and Artefacts: Or How the Sociology of Science and the Sociology of Technology Might Benefit Each Other," *Social Studies of Science* 14, no. 3 (1984): 399–441; and Roland Kline and Trevor Pinch, "Users as Agents of Technological Change: The Social Construction of the Automobile in the Rural United States," *Technology and Culture* 37, no. 4 (1996): 763–795.

9. Thomas Haigh, "Masculinity and the Machine Man: Gender in the History of Data Processing," in *Gender Codes: Why Women Are Leaving Computing*, ed. Thomas J. Misa (Los Alamitos, CA: Wiley-IEEE Computer Society Press, 2010), 51–72.

10. Frank Veraart, "Losing Meanings: Computer Games in Dutch Domestic Use, 1975–2000," *IEEE Annals of the History of Computing*, January–March 2011, 52–65.

11. IBM, "The IBM 700 Series: Computing Comes to Business," http://www-03 .ibm.com/ibm/history/ibm100/us/en/icons/ibm700series/impacts/ (accessed July 6, 2016).

12. Claude E. Shannon, "Programming a Computer for Playing Chess," *Philosophical Magazine* 41, no. 314 (1950). This article is posted online under http://archive .computerhistory.org/projects/chess/related_materials/text/ (accessed July 6, 2016) with pages numbered 1–18. The quotation is from page 1.

13. Jonathan Schaeffer, *One Jump Ahead: Challenging Human Supremacy in Checkers* (New York: Springer-Verlag, 1997), 90–102.

14. Alex Bernstein and Michael de V. Roberts, "Computer v. Chess Player," *Scientific American* 198, no. 6 (1958): 96–105.

15. The booklet accompanying Nim is posted online at "Nimrod," http://www .goodeveca.net/nimrod/NIMG_1.html#intro (accessed April 8, 2016).

16. Paul Ceruzzi, "From Scientific Instrument to Everyday Appliance: The Emergence of Personal Computers, 1970–1977," *History and Technology* 13 (1996): 1–31.

17. D. J. Edwards and J. M. Graetz, "PDP-1 Plays at Spacewar," *Decuscope* 1, no. 1 (1962): 2, 4.

18. Freiberger and Swaine, *Fire in the Valley*, 132.

19. "The Game of Life," *Time*, January 21, 1974, 76.

20. Freiberger and Swaine, *Fire in the Valley*, 132.

21. Leslie Haddon, "Electronic and Computer Games: The History of an Interactive Medium," *Screen* 29, no. 2 (1988): 52–75.

22. Donald D. Spencer, *Playing Games with Computers* (New York: Spartan Books, 1968).

23. Donald D. Spencer, "Game Playing with Computers," *Computers and Automation*, August 1968, 38–42.

24. *What to Do after You Hit Return, or P.C.C.'s First Book of Computer Games* (Menlo Park, CA: People's Computer Company, 1975).

25. Turner, *From Counterculture to Cyberculture*.

26. Timothy Ferris, "Solid State Fun," *Esquire*, March 1977, 101, 121–124.

27. See, e.g., "Digitizing," *New Yorker*, November 14, 1977, 40–42.

28. "Plugging in Everyman," *Time*, September 5, 1977, 43.

29. Ferris, "Solid State Fun."

30. Haddon, "The Home Computer"; Friedman, *Electric Dreams*, 96.

31. Thomas Streeter, *The Net Effect: Romanticism, Capitalism, and the Internet* (New York: NYU Press, 2011).

32. Friedman, *Electric Dreams*, 92–101.

33. Ibid.; Timothy Moy, "Culture, Technology, and the Cult of Tech in the 1970s," in *America in the Seventies*, ed. Beth Bailey and David Farber (Lawrence: University of Kansas Press, 2004), 208–228.

34. Edward K. Yasaki, Sr., "Microcomputers: For Fun and Profit?" *Datamation*, July 1977, 66–71.

35. Freiberger and Swaine, *Fire in the Valley*, 27.

36. Levy, *Hackers*, 243.

37. Nelson, *Computer Lib/Dream Machines*, is a challenging book to cite. It was self-published, underwent multiple editions and printings, and has separate pagination for each of its two halves. I am citing a volume described as "first edition" and "ninth printing," which was printed in 1983. I refer to page numbers as either *Computer Lib* (*CL*) or *Dream Machines* (*DM*).

38. *CL*, 2, 8.

39. *DM*, 2.

40. *DM*, 6.

41. *DM*, 22.

42. *DM*, 21.

43. *DM*, 21.

44. Turner, *From Counterculture to Cyberculture*.

45. Stewart Brand, "SPACEWAR: Fanatic Life and Symbolic Death among the Computer Bums," *Rolling Stone*, December 7, 1972. This article is available online at http://www.wheels.org/spacewar/stone/rolling_stone.html (accessed April 9, 2016). All subsequent quotations are from this online document.

46. Turner, *From Counterculture to Cyberculture*, 117.

47. Ceruzzi, *A History of Modern Computing*, 208.

48. Ferris, "Solid State Fun."

49. Freiberger and Swaine, *Fire in the Valley*, 133.

50. Levy, *Hackers*, 247–248.

51. Steve Wozniak with Gina Smith, *iWoz: How I Invented the Personal Computer, Co-founded Apple, and Had Fun Doing It* (New York: W. W. Norton, 2006), 140.

52. Friedman, *Electric Dreams*, 68–75, considers the representation of the computer in *2001*. On HAL's influence on science and culture see David G. Stork, ed., *HAL's Legacy: 2001's Computer as Dream and Reality* (Cambridge, MA: MIT Press, 1996).

53. *Time*, April 2, 1965; *Time*, February 20, 1978.

54. James T. Rogers, *The Calculating Book: Fun and Games with Your Pocket Calculator* (New York: Random House, 1975).

55. The trope of computer as thinking being was present in the early postwar years. See Edmund C. Berkeley, *Giant Brains or Machines That Think* (New York: Wiley, 1949).

56. Stephen Walton, "They Laughed When I Sat Down at the Computer...," *Popular Mechanics*, June 1978, 127–129.

57. David Welsh and Theresa Welsh, *Priming the Pump: How TRS-80 Enthusiasts Helped Spark the PC Revolution* (Ferndale, MI: The Seeker Book, 2007), 242.

58. *AudioVideo International*, November 1977, 23.

59. "And Now a Computer for Your Home," *Changing Times*, December 1977, 39–40.

60. Paul Trachtman, "A Generation Meets Computers on the Playing Fields of Atari," *Smithsonian*, September 1981, 50–61; 55.

61. Gary Hector, "Atari's New Game Plan," *Fortune*, August 8, 1983, 46–52.

62. "Atari's Struggle to Stay Ahead," *Business Week*, September 13, 1982, 56–60.

63. Some of these commercials might circulate online, but the versions I viewed were those in the collection of the UCLA Film & Television Archive, where the catalog gives the titles and agency.

5 Video Kids Endangered and Improved

1. Carolyn Marvin, *When Old Technologies Were New: Thinking about Electric Communication in the Late Nineteenth Century* (New York: Oxford University Press, 1988), 98.

2. Lynn Spigel, *Make Room for TV: Television and the Family Ideal in Postwar America* (Chicago: University of Chicago Press, 1992).

3. *City of Mesquite v. Aladdin's Castle, Inc.*, 455 US 283 (1982). The Dallas, Texas suburb barred children under seventeen from arcades if not accompanied by a parent or guardian, and also refused Aladdin's Castle Inc. a license to open one of its establishments in a shopping mall, contending that Aladdin's parent company, Bally, had ties to "criminal elements." A state court ordered that the arcade be allowed to open. The company brought a federal suit arguing that the "criminal element" law was unconstitutionally vague and that the regulation of patrons' ages violated their freedom of association. A federal appeals court found in favor of Aladdin's Castle; the Supreme Court ruled that the "criminal element" law was not too vague, but took no position on the age issue and sent the matter back to the appeals court. "High Court Sidesteps Video Game Battle Over Minors' Rights," *Wall Street Journal*, February 24, 1982, 52; Mary Claire Blakeman, "Courts Tackle Video Games: Education or Corruption?" *InfoWorld*, November 21, 1981, 24, 26, 29.

4. ABC's *Nightline*, February 26, 1983, begins with a clip of Robert Preston in *The Music Man*. In the introductory voice-over, the reporter begins, "These days the trouble begins with a V." Later in the episode, a concerned parent extends the River City analogy by asserting that video arcades are worse than the pool hall in *The Music Man* because of their ubiquity and their lack of rules. River City or the musical also appears in Blakeman, "Courts Tackle Video Games: Education or Corruption?"; William E. Geist, "The Battle for America's Youth: Long Island Mother Takes on Video Games," *New York Times*, January 5, 1982, B2; William A. Scrivener, "Video Arcades Are Stirring Up Main Street, N.J.," *New York Times*, May 23, 1982, New Jersey Weekly sec., 513; Charles Mount and Don Hayner, "Suburb's Ban on Video Games Is Ruled Illegal," *Chicago Tribune*, August 26, 1982, B3; Barbara Sullivan, "Judge Puts

Video Room Back on Track," *Chicago Tribune*, September 24, 1982, SD8; and "Donkey Kong Goes to Harvard," *Time*, June 6, 1983.

5. Kristen Drotner, "Dangerous Media? Panic Discourses and Dilemmas of Modernity," *Paedagogica Historica* 35, no. 3 (1999): 593–619; 609.

6. The distinction between moral panic and media panic is Drotner's in "Dangerous Media?" See also Angela McRobbie and Sarah Thornton, "Rethinking 'Moral Panic' for Multi-Mediated Social Worlds," *British Journal of Sociology* 46, no. 4 (1995): 559–574; John Springhall, *Youth, Popular Culture, and Moral Panics* (New York: St. Martin's Press, 1998); and Shayla Thiel-Stern, *From the Dance Hall to Facebook: Teen Girls, Mass Media, and Moral Panic* (Amherst: University of Massachusetts Press, 2014).

7. Staney Cohen, *Folk Devils and Moral Panics: The Creation of the Mods and Rockers*, 3rd ed. (London: Routledge, 2002).

8. Drotner, "Dangerous Media?," 615.

9. Ibid., 599.

10. Dmitri Williams, "Trouble in River City: The Social Life of Video Games," PhD diss., University of Michigan, 2004.

11. G. F. Cravenson, "Video Games for the 'Basest Instincts of Man,'" *New York Times*, January 28, 1982, A22.

12. Antonia van der Meer, "Video Games Can Be Hazardous to Your Health," *Mademoiselle*, July 1983, 61; Geist, "The Battle for America's Youth"; T. C. McCowan, "Space Invaders Wrist," *New England Journal of Medicine* 304 (1981): 1368; Ryan Pierson, "Making Sense of Early Video Arcades: The Case of Pittsburgh, 1980–1983," *Canadian Journal of Film Studies* 20, no. 2 (fall 2011): 19–37.

13. Desmond Ellis, "Video Arcades, Youth, and Trouble," *Youth and Society* 16, no. 1 (1984): 47–65; Nancy R. Needham, "Thirty Billion Quarters Can't Be Wrong—Or Can They?" *Today's Education* 71, no. 3 (1982): 52–55; Pam Reynolds, "Electronic Video Games: Friendly or Hostile Invasion?" *PTA Today*, December 1982–January 1983, 7–9.

14. Needham, "Thirty Billion Quarters Can't Be Wrong."

15. Scrivener, "Video Arcades Are Stirring Up Main Street," 513.

16. E.g., ABC News, "Video Games," December 9, 1981.

17. Mary Claire Blakeman, "Courts Tackle Video games: Education or Corruption?" *InfoWorld*, November 23, 1981, 24, 26, 29.

18. "Addictive Video Games," *Psychology Today*, May 1983, 87.

19. An Associated Press item ran as "Surgeon General Sees Danger in Video Games," *New York Times*, November 10, 1982, A16. The following day, the Surgeon General

released a statement clarifying that he had been expressing his personal opinion, not discussing any scientific evidence, and denied that video games are harmful to children. "Statement by Dr. E. Koop, Surgeon General," November 10, 1982, http://profiles.nlm.nih.gov/ps/access/QQBBCF.pdf (accessed April 8, 2016).

20. On the harms historically associated with television, see Michael Z. Newman and Elana Levine, *Legitimating Television: Media Convergence and Cultural Status* (New York: Routledge, 2011), 14–37.

21. Carly Kocurek discusses a number of these episodes in "Anarchy in the Arcade: Regulating Coin-Op Video Games," chapter 4 of *Coin-Operated Americans: Rebooting Boyhood at the Video Game Arcade* (Minneapolis: Minnesota University Press, 2015), 91–114.

22. See, e.g., Sue Mittenthal, "Video Games: Are They Harmful to Our Children? A Pro-Con Debate," *Family Circle*, July 20, 1982, 26–36.

23. "Moral entrepreneur" is a phrase taken from Howard S. Becker, *Outsiders: Studies in the Sociology of Deviance* (New York: Free Press of Glencoe, 1963); see also Thiel-Stern, *From the Dance Hall to Facebook*.

24. Harry Van Moorst, "Pinball and Moral Panics: Rumour, Allegation, and Ignorance as a Source of Media Sensationalism," Occasional Paper no. 3, Department of Humanities, Footscray Institute of Technology, Footscray, Victoria, 1980.

25. Mount and Hayner, "Suburb's Ban on Video Games Is Ruled Illegal," B3.

26. "Addictive Video Games."

27. Geist, "The Battle for America's Youth"; Glenn Collins, "Video Games: A Diversion or a Danger?" *New York Times*, February 17, 1983, C1; Fox Butterfield, "Video Game Specialists Come to Harvard to Praise Pac-Man, Not to Bury Him," *New York Times*, May 24 1983, A22.

28. Hilary de Vries, "Pow! Bang! Towns Zap Video Games," *Christian Science Monitor*, May 27, 1982; William Geist, "6-Foot-5 Pac-Man Man Is Scoring in Westport," *New York Times*, April 27, 1983, B3; Dudley Clendinen, "Massachusetts Town Exiles Pac-Man and All That," *New York Times*, December 8, 1983, A22; Pierson, "Making Sense of Early Video Arcades."

29. Reynolds, "Electronic Video Games." Reynolds also cites evidence from a study conducted by the New York State Division of Substance Abuse Services that drugs were used or sold in or near two-thirds of arcades in New York City.

30. Clendinen, "Massachusetts Town Exiles Pac-Man and All That."

31. Peter Kerr, "Should Video Games Be Restricted by Law?" *New York Times*, June 22, 1982, C1.

32. Collins, "Video Games."

33. Julius Segal and Zelda Segal, "Video Tripping: New Educational Video Games Just May Make It Worth the Ride," *Health*, June 1983, 24–29.

34. Sherry Turkle, *The Second Self: Computers and the Human Spirit* (New York: Simon & Schuster, 1985), 66.

35. Marie Winn, *The Plug-In Drug: Television, Children, and the Family* (New York: Viking, 1977).

36. I have transcribed this from a videotape of the program myself, eliminating vocalized pauses like "uh" but keeping the diction and syntax intact.

37. Howard Witt, "Debate over Video Arcade Storms into Morton Grove," *Chicago Tribune*, January 12, 1983, NS1.

38. "Town Torpedoes Plan for Mammoth Video Game Arcade," *Chicago Tribune*, February 18, 1983, NW4.

39. Robert E. Tomasson, "Westport Awaits Face-Off," *New York Times*, May 9, 1982, CN11.

40. Scrivener, "Video Arcades Are Stirring Up Main Street."

41. Geist, "6-Foot-5 Pac-Man Man Is Scoring in Westport."

42. John E. Sullivan, "Note: First Amendment Protection of Artistic Entertainment: Toward Reasonable Municipal Regulation of Video Games," *Vanderbilt Law Review* (October 1983): 1223; James D. Ivory, "Protecting Kids or Attacking the First Amendment? Video Games, Regulation, and Protected Expression," paper presented to the Law Division at the 86th annual conference of the Association for Education in Journalism and Mass Communication (AEJMC) Kansas City, MO, August 2003.

43. *America's Best Family Showplace Corp. v. City of New York*, Dept. of Bldgs., 536 F. Supp. 170 (E.D.N.Y. 1982).

44. *Malden Amusement Company, Inc. v. City of Malden*, 582 F. Supp. 297 (D. Mass. 1983).

45. David B. Goroff, "Note: The First Amendment Side Effects of Curing Pac-Man Fever," *Columbia Law Review* (April 1984): 744; Thomas Henry Rousse, "Electronic Games and the First Amendment: Free Speech Protection for New Media in the 21st Century," *Northwestern Interdisciplinary Law Review* 4, no. 1 (2011): 173.

46. The Supreme Court would recognize the status of games as free speech under the First Amendment only in 2011: *Brown, Governor of California v. Entertainment Merchants Association et al.*, 564 U.S. 786 (2011).

47. Chris Taylor, "It's On Like Donkey Kong: Town Repeals Arcade Game Ban," *Mashable*, May 2, 2014, http://mashable.com/2014/05/02/coin-op-ban-repealled/.

48. Howard Rheingold, "Video Games Go to School," *Psychology Today*, September 1983, 37–46; Carolyn Meinel, "Will Pac-Man Consume Our Nation's Youth?" *Technology Review* May–June 1983, 10–11, 28.

49. I am reading these experts in two ways, going in opposite directions. I am building on them but also thinking critically about their agenda. These experts are secondary sources, already analyzing evidence about computers and games and providing an interpretation of early 1980s American society and its technology. I rely on them for evidence that has already been interpreted, but is nonetheless still available for my own interpretation. I also see these experts as primary sources, evidence of the history of conventional wisdom and habits of thought (at least among highly educated members of society), and leaving behind traces of the everyday life of the time. So when one such figure refers to video games as a "futuristic technology packaged as home entertainment" (Mitchell in *Marriage and Family Review*, 1985, cited in n. 71 of this chapter), we can read this as a discourse in its own right, a fragment of historical data, rather than just a form of scholarly literature presenting an understanding based on empirical findings. While I am sympathetic to the experts defending video games against the crusading moral entrepreneurs, I am also skeptical of their optimistic technofuturism, and on the whole regard them with ambivalence.

50. Paul Trachtman, "A Generation Meets Computers on the Playing Field of Atari," *Smithsonian*, September 1981, 50–61. Carly Kocurek's history of early video arcades and masculinity echoes this notion. She argues that arcade games "introduced a generation of young men to computers as approachable, everyday technologies, just as the workplace was entering a period of massive computerization." Kocurek, *Coin-Operated Americans*, 12.

51. Daniel Bell, *The Coming of Post-Industrial Society: A Venture in Social Forecasting* (New York: Basic Books, 1973); Zbigniew Brzezinski, *Between Two Ages: America's Role in the Technetronic Era* (New York: Viking, 1970).

52. F. M. Esfandiary, *Telespheres* (New York: Popular Library, 1977).

53. Bell, *The Coming of Post-Industrial Society*, 125.

54. Ibid., 127.

55. Examples of popular books in addition to those cited above include Alvin Toffler, *The Third Wave* (New York: William Morrow, 1980) and John Naisbitt, *Megatrends: Ten New Directions Transforming Our Lives* (New York: Warner Books, 1982).

56. Matthew L. Wald, "Building for a Surging Service Economy," *New York Times*, May 15, 1983, 615.

57. Frank Webster, "Making Sense of the Information Age," *Information, Communication, and Society* 8, no. 4 (2005), doi:10.1080/13691180500418212 (accessed April 2, 2016).

58. For instance, a print advertisement for Apple Computers appearing in 1981 contained the following copy: "Putting real computer power in the hands of the individual is already improving the way people work, think, learn, communicate and spend their leisure hours. Computer literacy is fast becoming as fundamental a skill as reading or writing."

59. Kay Bremker, "Information Moves Ahead of Industry," *Washington Post*, September 25, 1983, AS76.

60. Everett M. Rogers, "The Diffusion of Home Computers among Households in Silicon Valley," *Marriage and Family Review* 8 (spring 1985): 89–102. This issue of *Marriage and Family Review* is on the theme of "Personal Computers and the Family" and contains many chapters that express the hopeful idea that computers in the home will benefit children.

61. Mitchell Robin, "Video Games Offer a Lot More Than Fun," *New York Times*, January 20, 1982, A26.

62. Joseph Psotka, "Computers and Education," *Behavior Research and Methods Instrumentation* 14 (1982): 221–223.

63. Rheingold, "Video Games Go to School."

64. Ellen Ruppel Shell, "Games People Program," *Technology Review*, November–December 1980, 10–13.

65. Thomas Malone, "What Makes Computer Games Fun?" in *CHI '81 Proceedings of the Joint Conference on Easier and More Productive Use of Computer Systems (Part—II): Human Interface and the User Interface* (New York: Association for Computer Machinery, 1981), 143.

66. Empirical research conducted at this time showed that there was no evidence in support of the hypothesis that playing games improves one's eye-hand coordination: Jerry L. Griffith et al., "Differences in Eye-Hand Motor Coordination of Video-Game Users and Non-Users," *Perceptual and Motor Skills* 57 (1983): 155–158.

67. David Surrey, "It's, Like, Good Training for Life," *Natural History*, November 1982, 71–83.

68. Rheingold, "Video Games Go to School"; Piaget is also central to the ideas in Seymour Papert, *Mindstorms: Children, Computers, and Powerful Ideas* (New York: Basic Books, 1980).

69. Isaac Asimov, "Video Games Are Dead, Long Live the Supergames of Tomorrow," *Video Review*, May 1983, 32–34.

70. Eric Levin, "They Zap, Crackle and Pop, but Video Games Can Be Powerful Tools for Learning," *People*, May 31, 1982.

71. Edna Mitchell, "The Dynamics of Family Interaction around Home Video Games," *Marriage and Family Review* 8 (spring 1985): 121–135.

72. Isaac Asimov, "The New Learning," *Videogaming Illustrated*, October 1982, 15–18.

73. Trachtman, "A Generation Meets Computers on the Playing Field of Atari," 56.

74. Ibid.

75. In addition to Geoffrey R. Loftus and Elizabeth F. Loftus, *Mind at Play: The Psychology of Video Games* (New York: Basic Books, 1983), popular nonfiction titles about video games by academic writers from this period include David Sudnow, *Pilgrim in the Microworld* (New York: Grand Central, 1979); Patricia Marks Greenfield, *Mind and Media: The Effects of Television, Video Games, and Computers* (Cambridge, MA: Harvard University Press, 1984); Papert, *Mindstorms*; and Turkle, *The Second Self*.

76. Loftus and Loftus, *Mind at Play*, x.

77. Ibid., 122–125.

78. Ibid., 124–25. See also Greenfield, *Mind and Media*, 105.

79. Ibid., 124–125.

80. Carolyn A. Yordan, "Behind the Screens: Teens Earn Big Bucks Designing Video Games," *Seventeen*, July 1983, 69.

81. *Video Games and Human Development: A Research Agenda for the '80s, Papers and Proceedings of a Symposium Held at the Harvard Graduate School of Education*, Cambridge, MA, May 22–24, 1983 (Cambridge, MA: Monroe C. Gutman Library, Harvard School of Education, 1983).

82. "Donkey Kong Goes to Harvard," *Time*, June 6, 1983; Charles Leerhsen, Marsha Zabarsky, and Dianne H. McDonald, "Video Games Zap Harvard," *Newsweek*, June 6, 1983, 92.

83. Fox Butterfield, "Video Game Specialists Come to Harvard to Praise Pac-Man," A22. Barbara Wierzbicki, "Video Arcades Meet Stiff Community Opposition," *InfoWorld*, December 26, 1983, 18–20, uses evidence presented at the Harvard conference in rebuttal of claims made by opponents of arcades.

84. *Video Games and Human Development*, 9.

85. Ibid., 13.

86. Ibid., 19–24.

87. Ibid., 30.

88. Ibid., 33–40.

89. Ibid., 34.

90. Thomas W. Malone, "What Makes Things Fun to Learn? A Study of Intrinsically Motivating Computer Games," Cognitive and Instructional Sciences Series CIS-7 (SSL-80-11), Xerox Palo Alto Research Center, August 1980.

91. *Video Games and Human Development*, 54.

92. Ibid., 4–7.

93. Ibid., 5.

94. Ibid., 25.

95. Ibid., 27.

96. Ibid., 14–16. This portion of the conference was publicized in popular press articles such as Butterfield, "Video Game Specialists Come to Harvard to Praise Pac-Man," and Wierzbicki, "Video Arcades Meet Stiff Community Opposition."

97. Ibid., 16.

98. Ibid., 62.

99. Ibid., 62–63.

100. McDonald, "Video Games Zap Harvard."

101. Kocurek, *Coin-Operated Americans*, 141, 144.

6   Pac-Man Fever

1. "Pac-Man Fever" was the 1981 single from the 1982 album of the same name by Buckner & Garcia. The song reached number 9 on the Billboard Hot 100 chart in March 1982, and became a gold record. (I have found no usage of the phrase "Pac-Man Fever" before the release of this record.)

2. Sherry Turkle, *The Second Self: Computers and the Human Spirit* (New York: Simon & Schuster, 1984), 66.

3. For a much more detailed description of *Pac-Man*'s design, see Jamey Pittman, "The Pac-Man Dossier," version 1.0.26, June 16, 2011, http://home.comcast.net/~jpittman2/pacman/pacmandossier.html (accessed June 30, 2015).

4. This phrase is often called "Bushnell's Law" after Nolan Bushnell. Ian Bogost, "Persuasive Games: Familiarity, Habituation, and Catchiness," *Gamasutra*, April 2, 2009, http://www.gamasutra.com/view/feature/3977/persuasive_games_familiarity (accessed April 9, 2016).

5. Chris Kohler, "Q&A: Pac-Man Creator Reflects on 30 Years of Dot-Eating," *Wired*, May 21, 2010, http://www.wired.com/2010/05/pac-man-30-years/ (accessed April 2,

2016). See also Steven L. Kent, *The Ultimate History of Video Games* (New York: Prima Publishing, 2001), 140–144.

6. Tristan Donovan, *Replay: The History of Video Games* (East Sussex, UK: Yellow Ant, 2010), 87.

7. Daniel Cohen, *Video Games* (New York: Archway, 1982); Rick Johnson, "What's Round and Yellow and Laughs All the Way to the Bank?" *Vidiot*, February–March 1983, 20–23; Mike Moore, "Videogames: Son of Pong," *Film Comment* January–February 1983, 34–37, 22; Chris Green, "Pac-Man," *Salon*, June 17, 2002, http://www.salon.com/2002/06/17/pac_man/ (accessed April 2, 2016).

8. Nick Montfort and Ian Bogost, *Racing the Beam: The Atari Video Computer System* (Cambridge, MA: MIT Press, 2009), 65–66; Jamey Pittman, "The Pac-Man Dossier," *Gamasutra*, http://www.gamasutra.com/view/feature/3938/the_pacman_dossier.php ?print=1 (accessed April 2, 2015).

9. "Pac-Man Fever," *Time*, April 5, 1982.

10. Scott Cohen, *Zap! The Rise and Fall of Atari* (Bloomington, IN: Xlibris, 1984), 95.

11. National Pac-Man Day was advertised in big-city newspapers around the United States. On the Atari cartridge's shortcomings, see Montfort and Bogost, *Racing the Beam*, 65–79.

12. Randy Fromm, "Pac-Attack! Pac-Attack!" *PlayMeter*, August 15, 1982, 48–49; Mike Shaw, "Video Rules the World's Fair," *PlayMeter*, August 15, 1982, 51–52.

13. Jennifer Smith, "Ms. Pac-Man," *Working Woman*, November 1982, 103. Mike Moore makes a similar point in "Videogames: Son of Pong."

14. Jennifer Allen, "All the World's a Video Game," *New York*, January 17, 1983.

15. Sidney J. Kaplan and William Beckham, "Videos: Differences between Male and Female Players," *PlayMeter*, February 15, 1982, 28–31.

16. Marion Cutler and Jane Petersson, "Do Coin-Ops Shut Out Potential Distaff Fans?" *PlayMeter*, January 1, 1982, 37–38.

17. Mary Claire Blakeman, "The Video Games Women Play … And Why," *PlayMeter* May 1, 1982, 34–38.

18. Ibid.

19. Consumer Guide Editors, *How to Win at Video Games* (New York: Simon & Schuster, 1982), 82–89.

20. Daniel Cohen, *Video Games* (New York: Archway, 1982).

21. Ibid.

22. "The Signal on the Screen," *Vidiot*, September–October 1982, 4–5.

23. "What's Round and Yellow and Laughs all the Way to the Bank?" *Vidiot*, February–March 1983, 20–23.

24. Rick Johnson, "Twerp-Factor Up!" *Vidiot*, February/–March 1983, 22.

25. Mark Baker, *I Hate Vidiots: Today the Arcade, Tomorrow the World* (New York: Fireside Books, 1982), 52.

26. Ibid., 49.

27. Ibid., 55.

28. Ibid., 53–54.

29. "Video Game Ruling," *New York Times*, March 19, 1982, D5.

30. Doug Macrae, presentation at California Extreme 2010, https://vimeo.com/15532555 (accessed June 29, 2015); Kent, *The Ultimate History of Video Games*, 172.

31. "Here's the Queen of the Video Game Scene!" *Electronic Games*, July 1983, 38.

32. The Arcade Flyer Museum, http://flyers.arcade-museum.com/flyers_video/midway/62503001.jpg (accessed June 29, 2015).

33. "Ms. Pac-Man," *Vidiot*, August–September 1983, 39.

34. "Ms. Pac-Man," *Blip*, August 1983, 30–31.

35. "The Machines," *New Yorker*, October 4, 1982, 33–35.

36. "Pac Man Fever," *Square Pegs*, originally aired October 11, 1982. It was not unheard of for schools to have video games in game rooms at this time; one example was Nicolet High School in Glendale, Wisconsin, referenced in Mary Claire Blakeman, "Why Parents Want to Restrict Video Game Play—and What the Industry Can Do," *PlayMeter*, December 1, 1981, 102–103.

Select Bibliography

Acland, Charles R. *Swift Viewing: The Popular Life of Subliminal Influence*. Durham: Duke University Press, 2012.

Adorno, Theodor W., and Max Horkheimer. "The Culture Industry: Enlightenment as Mass Deception." In *Dialectic of Enlightenment: Philosophical Fragments*, trans. Edmund Jephcott. Palo Alto, CA: Stanford University Press, 1944/2002.

Anderson, Tim. "New Media? Please Define." *Flow*, May 12, 2006, http://www.flowjournal.org/2006/05/new-media-please-define/.

Baer, Ralph H. *Videogames in the Beginning*. Springfield, NJ: Rolenta Press, 2005.

Baker, Mark. *I Hate Vidiots: Today the Arcade, Tomorrow the World*. New York: Fireside Books, 1982.

Baym, Nancy. *Personal Connections in the Digital Age*. London: Polity, 2010.

Becker, Howard S. *Outsiders: Studies in the Sociology of Deviance*. New York: Free Press of Glencoe, 1963.

Bell, Daniel. *The Coming of Post-Industrial Society: A Venture in Social Forecasting*. New York: Basic Books, 1973.

Bell, Daniel. "The Theory of Mass Society: A Critique." In *Mass Communication and American Social Thought: Key Texts 1919–1968*, ed. John Durham Peters and Peter Simonson, 364–373. Lanham: Rowman & Littlefield, 2014.

Berkeley, Edmund C. *Giant Brains or Machines that Think*. New York: Wiley, 1949.

Berker, T., M. Hartmann, Y. Punie, and K. Ward, eds. *Domestication of Media and Technology*. Maidenhead, UK: Open University Press, 2006.

Bernstein, Alex, and Michael de V. Roberts. "Computer v. Chess Player," *Scientific American* 198, no. 6 (1958): 96–105.

Boddy, William. *Fifties Television: The Industry and Its Critics*. Champaign, IL: University of Illinois Press, 1990.

Boddy, William. *New Media and Popular Imagination: Launching Radio, Television, and Digital Media in the United States*. New York: Oxford University Press, 2004.

Bogost, Ian. "Persuasive Games: Familiarity, Habituation, and Catchiness." *Gamasutra*, April 2, 2009, http://www.gamasutra.com/view/feature/3977/persuasive_games_familiarity.

Bolter, David Jay, and Richard Grusin. *Remediation: Understanding New Media*. Cambridge, MA: MIT Press, 1999.

Brand, Stewart. "SPACEWAR: Fanatic Life and Symbolic Death among the Computer Bums." *Rolling Stone*, December 7, 1972.

Brzezinski, Zbigniew. *Between Two Ages: America's Role in the Technetronic Era*. New York: Viking, 1970.

Burnham, Van. *Supercade: A Visual History of the Videogame Age, 1971–1983*. Cambridge, MA: MIT Press, 2001.

Campbell-Kelly, Martin. *From Airline Reservations to Sonic the Hedgehog: A History of the Software Industry*. Cambridge, MA: MIT Press, 2003.

Ceruzzi, Paul E. "From Scientific Instrument to Everyday Appliance: The Emergence of Personal Computers, 1970–1977." *History and Technology* 13 (1996): 1–31.

Ceruzzi, Paul E. *A History of Modern Computing*, 2nd ed. Cambridge, MA: MIT Press, 2003.

Clark, Alison J. "Coming of Age in Suburbia: Gifting the Consumer Child." In *Designing Modern Childhoods: History, Space, and the Material Culture of Children*, ed. Marta Gutman and Ning de Coninck-Smith, 253–268. New Brunswick, NJ: Rutgers University Press, 2008.

Clark, Clifford J., Jr. *American Family Home, 1800–1960*. Chapel Hill: University of North Carolina Press, 1986.

Cohen, Daniel. *Video Games*. New York: Archway, 1982.

Cohen, Scott. *Zap! The Rise and Fall of Atari*. Bloomington, IN: Xlibris, 1984.

Cohen, Stanley. *Folk Devils and Moral Panics: The Creation of the Mods and Rockers*, 3rd ed. London: Routledge, 2002.

Consumer Guide Editors. *How to Win at Video Games*. New York: Simon & Schuster, 1982.

Cook, David A. *Lost Illusions: American Cinema in the Shadow of Vietnam and Watergate, 1970–1979*. New York: Scribner's, 2000.

Czitrom, Daniel J. *Media and the American Mind: From Morse to McLuhan*. Chapel Hill: University of North Carolina Press, 1982.

Davis, Michael M., Jr. *The Exploitation of Pleasure: A Study of Commercial Recreations in New York City*. New York: Department of Child Hygiene of the Russell Sage Foundation, 1911.

Dawson, Max. "Home Video and the 'TV Problem': Cultural Critics and Technological Change." *Technology and Culture* 48 (2007): 524–549.

Dennett, Andrea Stulman. *Weird and Wonderful: The Dime Museum in America*. New York: NYU Press, 1997.

Dillon, Roberto. *The Golden Age of Video Games: The Birth of a Multi-Billion Dollar Industry*. Boca Raton, FL: CRC Press, 2011.

Donovan, Tristan. *Replay: The History of Video Games*. East Sussex, UK: Yellow Ant, 2010.

Drotner, Kristen. "Dangerous Media? Panic Discourses and Dilemmas of Modernity." *Paedagogica Historica* 35, no. 3 (1999): 593–619.

Ellis, Desmond. "Video Arcades, Youth, and Trouble." *Youth and Society* 16, no. 1 (1984): 47–65.

Elsaesser, T., ed. *Early Cinema: Space Frame Narrative*. London: BFI, 1990.

Esfandiary, F. M. *Telespheres*. Popular Library, 1977.

Fischer, Claude S. *America Calling: A Social History of the Telephone to 1940*. Berkeley: University of California Press, 1992.

Flynn, Bernadette. "Geographies of the Digital Hearth." *Information Communication and Society* 6, no. 4 (2003): 551–576.

Freiberger, Paul, and Michael Swaine. *Fire in the Valley: The Making of the Personal Computer*. Berkeley, CA: Osborne/McGraw-Hill, 1984.

Friedman, Ted. *Electric Dreams: Computers in American Culture*. New York: NYU Press, 2005.

Gitelman, Lisa. *Always Already New: Media, History and the Data of Culture*. Cambridge, MA: MIT Press, 2006.

Goroff, David B. "Note: The First Amendment Side Effects of Curing Pac-Man Fever." *Columbia Law Review* 84, no. 3 (April 1984): 744–774.

Goss, Jon. "The 'Magic of the Mall': An Analysis of Form, Function, and Meaning in the Contemporary Retail Built Environment." *Annals of the Association of American Geographers* 83, no. 1 (1993): 18–47.

Green, Chris. "Pac-Man." *Salon*, June 17, 2002, http://www.salon.com/2002/06/17/pac_man/.

Greenfield, Patricia Marks. *Mind and Media: The Effects of Television, Video Games, and Computers*. Cambridge, MA: Harvard University Press, 1984.

Griffith, Jerry L., Patricia Voloschin, Gerald D. Gibb, and James R. Bailey. "Differences in Eye-Hand Motor Coordination of Video-Game Users and Non-Users." *Perceptual and Motor Skills* 57 (1983): 155–158.

Gruen, Victor, and Larry Smith. *Shopping Towns USA: The Planning of Shopping Centers*. New York: Reinhold, 1960.

Guins, Raiford. *Game After: A Cultural Study of Video Game Afterlife*. Cambridge, MA: MIT Press, 2014.

Gunning, Tom. "The Cinema of Attraction: Early Film, Its Spectator, and the Avant-Garde." *Wide Angle* 8 (1986): 63–70.

Gutman, M., and N. de Coninck-Smith, eds. *Designing Modern Childhoods: History, Space and the Material Culture of Children*. New Brunswick, NJ: Rutgers University Press, 2008.

Haddon, Leslie. "Electronic and Computer Games: The History of an Interactive Medium." *Screen* 29, no. 2 (1988): 52–75.

Haddon, Leslie. "The Home Computer: The Making of a Consumer Electronic." *Science as Culture* 2 (1988): 7–51.

Haigh, Thomas. "Masculinity and the Machine Man: Gender in the History of Data Processing." In *Gender Codes: Why Women are Leaving Computing*, ed. Thomas J. Misa, 51–72. Los Alamitos, CA: Wiley-IEEE Computer Society Press, 2010.

Harris, Neil. "Spaced Out at the Shopping Center." In his *Cultural Excursions: Marketing Appetites and Cultural Tastes in Modern America*, 278–288. Chicago: University of Chicago Press, 1990.

Hartt, Rollin Lynde. *The People at Play*. Boston: Houghton Mifflin, 1909.

Herman, Leonard. "Ball-and-Paddle Consoles." In *Before the Crash: Early Video Game History*, ed. Mark J. P. Wolf, 53–59. Detroit: Wayne State University Press, 2012.

Huhtamo, Erkii. "Slots of Fun, Slots of Trouble: An Archaeology of Arcade Gaming." In *Handbook of Computer Game Studies*, ed. Joost Raessens and Jeffrey Goldstein, 3–21. Cambridge, MA: MIT Press, 2005.

Huhtamo, Erkki. "What's Victoria Got to Do with It? Toward an Archaeology of Domestic Video Gaming." In *Before the Crash: Early Video Game History*, ed. Mark J. P. Wolf, 30–52. Detroit: Wayne State University Press, 2012.

Hunt, Andrew. "How New Was the New Left? Re-Thinking New Left Exceptionalism." In *The New Left Revisited*, ed. John McMillian and Paul Buhle, 139–155. Philadelphia: Temple University Press, 2003.

Ivory, James D. "Protecting Kids or Attacking the First Amendment? Video Games, Regulation, and Protected Expression." Paper presented to the Law Division at the 86th Annual Conference of the Association for Education in Journalism and Mass Communication (AEJMC), Kansas City, MO, August 2003.

Jackson, Kenneth. *Crabgrass Frontier: The Suburbanization of the United States*. New York: Oxford University Press, 1987.

Jacobson, Lisa. *Raising Consumers: Children and the American Mass Market in the Early Twentieth Century*. New York: Columbia University Press, 2005.

Jauss, Hans Robert, and Elizabeth Benzinger. "Literary History as a Challenge to Literary Theory." *New Literary History* 2, no. 1 (1970): 7–37.

Jenkins, Henry. "Complete Freedom of Movement: Video Games as Gendered Play Spaces." In *The Game Design Reader: A Rules of Play Anthology*, ed. Katie Salen and Eric Zimmerman, 330–363. Cambridge, MA: MIT Press, 2006.

Jenkins, Henry. *Convergence Culture: Where Old and New Media Collide*. New York: New York University Press, 2006.

Kaplan, Sidney J. "The Image of Amusement Arcades and Differences in Male and Female Video Game Playing." Paper presented at the 1982 annual meeting of the North Central Sociological Association.

Kent, Steven L. *The Ultimate History of Video Games*. Roseville, CA: Prima Publishing, 2001.

Kline, Roald, and Trevor Pinch. "Users as Agents of Technological Change: The Social Construction of the Automobile in the Rural United States." *Technology and Culture* 37, no. 4 (1996): 763–795.

Kocurek, Carly. "Coin-Drop Capitalism: Economic Lessons from the Video Arcade." In *Before the Crash: Early Video Game History*, ed. Mark J. P. Wolf, 189–208. Detroit: Wayne State University Press, 2012.

Kocurek, Carly. *Coin-Operated Americans: Rebooting Boyhood at the Video Game Arcade*. Minneapolis: University of Minnesota Press, 2015.

Kohler, Chris. "Q&A: Pac-Man Creator Reflects on 30 Years of Dot-Eating." *Wired*, May 21, 2010, http://www.wired.com/2010/05/pac-man-30-years/.

Kowinski, William Severini. *The Malling of America: An Inside Look at the Great Consumer Paradise*. New York: William Morrow, 1985.

Levy, Steven. *Hackers: Heroes of the Computer Revolution*. New York: Anchor, 1984.

Loftus, Geoffrey R., and Elizabeth F. Loftus. *Mind at Play: The Psychology of Video Games*. New York: Basic Books, 1983.

Lowood, Henry. "Videogames in Computer Space: The Complex History of Pong." *IEEE Annals of the History of Computing* 31, no. 1 (2009): 5–19.

Maddox, Brenda. *Beyond Babel: New Directions in Communication*. New York: Simon & Schuster, 1972.

Malone, Thomas. "What Makes Computer Games Fun?" In *CHI '81 Proceedings of the Joint Conference on Easier and More Productive Use of Computer Systems (Part II): Human Interface and the User Interface*, volume 1981, 143. New York: ACM, 1981.

Malone, Thomas W. "What Makes Things Fun to Learn? A Study of Intrinsically Motivating Computer Games." Cognitive and Instructional Sciences Series CIS-7 (SSL-80-11), Xerox Palo Alto Research Center, August 1980.

Manovich, Lev. *The Language of New Media*. Cambridge, MA: MIT Press, 2002.

Markoff, John. *What the Dormouse Said: How the Sixties Counterculture Shaped the Personal Computer Industry*. New York: Viking, 2005.

Martin, Michèle. "The Culture of the Telephone." In *Sex/Machine: Readings in Culture, Gender and Technology*, ed. Patrick D. Hopkins, 50–74. Bloomington, IN: Indiana University Press, 1989.

Marvin, Carolyn. *When Old Technologies Were New: Thinking about Electric Communication in the Late Nineteenth Century*. New York: Oxford University Press, 1988.

McDonald, Christopher Felix. "Building the Information Society: A History of Computing as a Mass Medium." PhD diss., Princeton University, 2011.

McNamara, Brooks. "'A Congress of Wonders': The Rise and Fall of Dime Museums." *ESQ: Journal of the American Renaissance* 20, no. 3 (1974): 216–232.

McRobbie, Angela, and Sarah Thornton. "Rethinking 'Moral Panic' for Multi-Mediated Social Worlds." *British Journal of Sociology* 46, no. 4 (1995): 559–574.

Mills, C. Wright. "Letter to the New Left." *New Left Review* 5 (1960).

Mills, C. Wright. "The Mass Society." In *Mass Communication and American Social Thought: Key Texts 1919–1968*, ed. John Durham Peters and Peter Simonson, 387–400. Lanham: Rowman & Littlefield, 2014.

Mills, C. Wright. *The Power Elite*. New York: Oxford University Press, 1956.

Mintz, Steven. *Huck's Raft: A History of American Childhood*. Cambridge, MA: Belknap Press of Harvard University Press, 2006.

Mitchell, Edna. "The Dynamics of Family Interaction around Home Video Games." *Marriage and Family Review* 8 (1985): 121–135.

Montfort, Nick, and Ian Bogost. *Racing the Beam: The Atari Video Computer System.* Cambridge, MA: MIT Press, 2009.

van Moorst, Harry. "Pinball and Moral Panics: Rumour, Allegation and Ignorance as a Source of Media Sensationalism." Occasional Paper no. 3, Department of Humanities, Footscray Institute of Technology, Footscray, Victoria, 1980.

Moy, Timothy. "Culture, Technology, and the Cult of Tech in the 1970s." In *America in the Seventies*, ed. Beth Bailey and David Farber. 208–228. Lawrence, KS: University of Kansas Press, 2004.

Muller, Peter O. *Contemporary Suburban America.* Englewood Cliffs, NJ: Prentice-Hall, 1981.

Murphy, Kay Rohl. "Family Patterns of Use and Parental Attitudes toward Home Electronic Video Games and Future Technology." PhD diss., Oklahoma State University, 1984.

Murphy, Sheila C. *How Television Invented New Media.* New Brunswick, NJ: Rutgers University Press, 2011.

Musser, Charles. *The Emergence of Cinema: The American Screen to 1907.* Berkeley: University of California Press, 1990.

Naisbitt, John. *Megatrends: Ten New Directions Transforming Our Lives.* New York: Warner Books, 1982.

Nasaw, David. *Going Out: The Rise and Fall of Public Amusements.* New York: Basic Books, 1993.

Needham, Nancy R. "Thirty Billion Quarters Can't Be Wrong—Or Can They?" *Today's Education* 71, no. 3 (1982): 52–55.

Nelson, George. *Tomorrow's House: A Complete Guide for the Home-Builder.* New York: Simon & Schuster, 1945.

Nelson, Ted. *Computer Lib/Dream Machines.* Self-published, 1974.

Neuman, W. Russell. *The Future of the Mass Audience.* Cambridge: Cambridge University Press, 1991.

Newman, Michael Z. "Free TV: File-Sharing and the Value of Television." *Television and New Media* 13 (2012): 463–479.

Newman, Michael Z. *Video Revolutions: On the History of a Medium.* New York: Columbia University Press, 2014.

Newman, Michael Z. "When Television Marries Computer." *Flow*, November 18, 2013, http://www.flowjournal.org/2013/11/when-television-marries-computer/.

Newman, Michael Z., and Elana Levine. *Legitimating Television: Media Convergence and Cultural Status*. New York: Routledge, 2012.

Papert, Seymour. *Mindstorms: Children, Computers, and Powerful Ideas*. New York: Basic Books, 1980.

Peters, Benjamin. "Lead Us not into Thinking the New Is New: A Bibliographic Case for New Media History." *New Media and Society* 11 (2009): 13–30.

Pierson, Rya. "Making Sense of Early Video Arcades: The Case of Pittsburgh, 1980–1983." *Canadian Journal of Film Studies* 20, no. 2 (2011): 19–37.

Pinch, Trevor J., and Wiebe E. Bijker. "The Social Construction of Facts and Artefacts: Or How the Sociology of Science and the Sociology of Technology Might Benefit Each Other." *Social Studies of Science* 14 (1984): 399–441.

Pittman, Jamey. "The Pac-Man Dossier." *Gamasutra*, http://www.gamasutra.com/view/feature/3938/the_pacman_dossier.php?print=1. Accessed April 2, 2015.

Psotka, Joseph. "Computers and Education." *Behavior Research Methods and Instrumentation* 14 (1982): 221–223.

Riismandel, Kyle. "Arcade Addicts and Mallrats: Producing and Policing Suburban Public Space in 1980s America." *Environment, Space, Place* 5, no. 2 (2013): 65–89.

Rogers, Everett M. "The Diffusion of Home Computers among Households in Silicon Valley." *Marriage and Family Review* 8 (spring 1985): 89–102.

Rosenberg, B., and D. M. White, eds. *Mass Culture: The Popular Arts in America*. Glencoe, IL: The Free Press, 1958.

Rotundo, E. Anthony. "Boy Culture." In *The Children's Culture Reader*, ed. Henry Jenkins, 337–362. New York: NYU Press, 1998.

Rousse, Thomas Henry. "Electronic Games and the First Amendment: Free Speech Protection for New Media in the 21st Century." *Northwestern Interdisciplinary Law Review* 4, no. 1 (2011): 173.

Rybczynski, Witold. *Home: A Short History of an Idea*. New York: Viking Penguin, 2001.

Schaeffer, Jonathan. *One Jump Ahead: Challenging Human Supremacy in Checkers*. New York: Springer, 1997.

Schlereth, Thomas J. *Victorian America: Transformations in Everyday Life, 1876–1915*. New York: HarperCollins, 1991.

Shannon, Claude E. "Programming a Computer for Playing Chess." *Philosophical Magazine* 41, no. 314 (1950).

Sharpe, Roger C. *Pinball!* New York: E. P. Dutton, 1977.

Shaw, Adrienne. "Do You Identify as a Gamer? Gender, Race, Sexuality, and Gamer Identity." *New Media and Society* 14 (2012): 28–44.

Shaw, Adrienne. "On not Becoming Gamers: Moving beyond the Constructed Audience." *Ada: A Journal of Gender, New Media, and Technology* 2 (2013). doi:10.7264/N33N21B3.

Shils, Edward. "Mass Society and Its Culture." *Deadalus* 89, no. 2 (1960): 288–314.

Smith, Ralph Lee. *The Wired Nation*. New York: Harper Colophon, 1972.

Spencer, Donald D. *Playing Games with Computers*. New York: Spartan Books, 1968.

Spigel, Lynn. *Make Room for TV: Television and the Family Ideal in Postwar America*. Chicago: University of Chicago Press, 1992.

Spigel, Lynn. *Welcome to the Dreamhouse: Popular Media and Postwar Suburbs*. Durham, NC: Duke University Press, 2001.

Springhall, John. *Youth Popular Culture, and Moral Panics: Penny Gaffs to Gangsta Rap, 1830–1996*. New York: St. Martin's Press, 1998.

Stork, David G., ed. *HAL's Legacy: 2001's Computer as Dream and Reality*. Cambridge, MA: MIT Press, 1996.

Streeter, Thomas. "Blue Skies and Strange Bedfellows: The Discourse of Cable Television." In *The Revolution Wasn't Televised: Sixties Television and Social Conflict*, ed. Lynn Spigel and Michael Curtin, 221–242. New York: Routledge, 1997.

Streeter, Thomas. *The Net Effect: Romanticism, Capitalism, and the Internet*. New York: NYU P, 2011.

Sturken, Marita. "TV as a Creative Medium: Howard Wise and Video Art." *AfterImage* (May 1984): 5–9.

Sudnow, David. *Pilgrim in the Microworld*. New York: Warner Books, 1983.

Sullivan, John E. "Note: First Amendment Protection of Artistic Entertainment: Toward Reasonable Municipal Regulation of Video Games." *Vanderbilt Law Review* 36, no. 5 (October 1983): 1223.

Sutton-Smith, Brian. *Toys as Culture*. Kidron, OH: Gardner Press, 1986.

Thiel-Stern, Shayla. *From the Dance Hall to Facebook: Teen Girls, Mass Media, and Moral Panic*. Amherst: University of Massachusetts Press, 2014.

Tichi, Cecilia. *Electronic Hearth: Creating an American Television Culture*. New York: Oxford University Press, 1992.

Toffler, Alvin. *The Third Wave*. New York: William Morrow, 1980.

Trapunski, Edward. *Special When Lit: A Visual and Anecdotal History of Pinball*. Garden City, NY: Dolphin Books, 1979.

Turkle, Sherry. *The Second Self: Computers and the Human Spirit*. New York: Simon & Schuster, 1984.

Turner, Fred. *The Democratic Surround: Multimedia and American Liberalism from World War II to the Psychedelic Sixties*. Chicago: University of Chicago Press, 2013.

Turner, Fred. *From Counterculture to Cyberculture: Stewart Brand, the Whole Earth Network, and the Rise of Digital Utopianism*. Chicago: University of Chicago Press, 2006.

Veraart, Frank. "Losing Meanings: Computer Games in Dutch Domestic Use, 1975–2000." *IEEE Annals of the History of Computing* 33, no. 1 (2011): 52–65.

Video Games and Human Development: A Research Agenda for the '80s, Papers and Proceedings of a Symposium Held at the Harvard Graduate School of Education, Cambridge, MA, May 22–24, 1983. Cambridge, MA: Monroe C. Gutman Library, Harvard School of Education, 1983.

Viorst, Milton. *Fire in the Streets: America in the 1960s*. New York: Simon & Schuster, 1979.

Watkins, Ralph J. "A Competitive Assessment of the U.S. Video Game Industry." United States International Trade Commission, Washington, DC, March 1984.

Webster, Frank. "Making Sense of the Information Age." *Information Communication and Society* 8 (2005): 4. doi:10.1080/13691180500418212.

Welsh, David, and Theresa Welsh. *Priming the Pump: How TRS-80 Enthusiasts Helped Spark the PC Revolution*. Ferndale, MI: The Seeker Book, 2007.

What to Do after You Hit Return, or P.C.C.'s First Book of Computer Games. Menlo Park, CA: People's Computer Company, 1975.

Williams, Dmitri. "Trouble in River City: The Social Life of Video Games." PhD diss., University of Michigan, 2004.

Williams, Dmitri. "The Video Game Lightning Rod: Constructions of a New Media Technology, 1970–2000." *Information Communication and Society* 6, no. 4 (2003): 523–550.

Williams, Raymond. *Television: Technology and Cultural Form*. Hanover, NH: Wesleyan University Press, 1974.

Wilson, Jason. "'Participation TV': Videogame Archaeology and New Media Art." In *The Pleasures of Computer Gaming: Essays on Cultural History, Theory, and Aesthetics*, ed. Melanie Swalwell and Jason Wilson, 94–117. Jefferson, NC: MacFarland, 2008.

Winn, Marie. *The Plug-In Drug: Television, Children, and the Family.* New York: Viking, 1977.

Wozniak, Steve, with Gina Smith. *iWoz: How I Invented the Personal Computer, Co-founded Apple, and Had Fun Doing It.* New York: W. W. Norton, 2006.

Index